HOOKED
BUT NOT
HELPLESS

KICKING NICOTINE ADDICTION

PATRICIA ALLISON

WITH JACK YOST

BridgeCity Books • Portland, Oregon

"Hooked—But Not Helpless:
Kicking Nicotine Addiction"
Third edition, 1999
Copyright © 1999, 1996 by Patricia Allison

Previously published as:
"Hooked—But Not Helpless:
Ending Your Love/Hate Relationship with Nicotine"
Copyright © 1990, 1992 by Patricia Allison

Published by:
BridgeCity Books
1717 SW Park, Suite 616, Portland, OR 97201
Info: 503-220-4171, Fax: 503-220-4081
Individual orders: 800-884-4171
Web page: www.stopsmoking.com

Distribution:
Independent Publishers Group
814 N. Franklin St., Chicago, IL 60610
Tel: 800-888-4741

Publisher's Cataloging Data:
Allison, Patricia L.
Hooked—But Not Helpless: Kicking Nicotine Addiction
1, Tobacco Addiction. 2, Self-help/Recovery. 3, Health. I. Title.
ISBN 0-9623683-7-7

Cover Design by Kari Powell
Author Photo by Scott Green
Illustrations by Elsa Warnick
Make Up by Gwen Hall

Made in the U.S.A.

LETTERS

"I have included *Hooked—But Not Helpless* in our cardiac rehabilitation program. Your book has helped many patients succeed who have failed in numerous other attempts to stop smoking, because it teaches them how to find the power within themselves to accomplish what needs to be done. I recommend it highly."

— David Bee, MD
Glendale, California

"I was a heavy smoker for 44 years, since I was sixteen, and I was one of those people who had tried everything and nothing worked. I was obsessed with stopping. I was in a state of panic, knowing it was harming me. I felt out of control, in conflict with all my other values. Oh, yes, I could stop smoking—but I always rationalized my way back to it. I thought I was special and could get away with it. Then a friend gave me your book. As I read it, you were always one step ahead of all my clever little schemes to make it okay to smoke, and I realized I was just a run-of-the-mill addict.

"Your book is brilliant. I finally found the answers to my problem. I was able to stop and haven't smoked since. I'm ordering two more books because I gave away my first one and now I want one just for security and the other to lend out."

— Virginia Barnett
Atlanta, Georgia

"In 1985 I saved my life by coming to your program. I had been told several times by my doctor to stop smoking due to blood clots. What she couldn't tell me was *how* to stop. You did, and then you stayed with it and talked me through the first year.

"I learned so much about myself and how to take control of my life with the techniques you teach. I have all the things I wanted from becoming an ex-smoker and so much more. Not only am I alive, but the quality of my life has improved 500%. I have freedom I never dreamed of before coming to you. Thank you so much for being here."

— Amanda Colorado
Portland, Oregon

CONTENTS

PREFACE

Smoking—drug addiction or bad habit?

Since *Hooked—But Not Helpless* was first published in 1990, important new evidence that tobacco is powerfully addictive has surfaced, capturing the media spotlight.

Recent studies prove not only that nicotine is addictive but that it is the *most addictive* of all drugs in important ways.

Two independent researchers, in major studies comparing nicotine, heroin, cocaine and alcohol, rank nicotine *first* in the power to create dependency among users. Included in the experts' definition of dependency was: "How difficult it is for the user to quit, the relapse rate, and the degree to which the substance is used in the face of evidence that it causes harm."

These findings only re-confirm earlier studies, including former Surgeon General C. Everett Koop's landmark 1988 report, which put nicotine in the same category as the most addictive illegal drugs. The few lonely voices still disputing these conclusions, not surprisingly, belong to the tobacco industry.

If so much is known about nicotine's addictive power, why do so many of the leading stop-smoking programs continue to treat smoking as something you should be able to just walk away from—with a little willpower, a few tips, and a word of encouragement?

Clearly, they still don't get it. There is a major chasm between *labeling* nicotine as a drug addiction and *treating* it as one.

The crux of the problem is this: although smoking *is* a drug addiction, it doesn't *look* like one. People don't associate smokers' behavior with that of drunks or dopers. Smoking cigarettes doesn't make people dangerous, tipsy, or out of control. And as long as cigarettes remain available, nobody's going to steal, kill or risk their lives to get their hands on them.

So despite all the research identifying nicotine as a drug dependency, in most people's minds smoking remains a bad habit. And it's treated like one. As the headline of the recent newspaper article from a major U.S. daily advised smokers: "Having a withdrawal pang? Take a walk, a brisk one."

But as I've been telling my clients for the last 16 years, "Don't imagine you can simply walk away from this, anymore than you could expect to walk away from heroin or alcohol addiction. No matter how often or how fast you walk around the block, guess what's going to be waiting for you when you get back to where you started?"

Chronic smokers have a deeply personal relationship with their cigarettes, one that usually goes all the back to their teenage years. This is obvious from the way they talk about them.

"I'd give anything for a cigarette."

"Cigarettes are my best friend."

"My cigarettes are always there for me."

Even if you love your cigarettes, you probably hate what they are doing to you. You want to smoke *and* you want to stop smoking—which leaves you in a terrible bind.

In this book, I do not promise you a quick-and-easy cure, but you will discover here a lucid and thorough understanding of the problem of nicotine addiction. You will learn how to stop smoking without driving yourself crazy or gaining a lot of weight. And you will learn how to defend yourself against the tricks your own mind will play to get you back to smoking.

You don't have to be ready to stop smoking to read this book. Go ahead, smoke—but keep reading. You won't find any blame or guilt here, only the information you need to stop smoking when you're ready.

This book is not meant to take the place of a good stop-smoking program or the extra support and guidance you may need from a physician, self-help group, or counseling . But it could well be your key to freedom, the help you need to finally end your love-hate relationship with nicotine.

INTRODUCTION

A FIGHTING CHANCE

Carrot sticks, deep breathing, a walk around the block. Now that you know what doesn't work—it's time to take a serious look at something that does.

If you've been fighting hard to stop smoking and are *still* losing the battle, there's something you need to know. Smoking is a drug addiction, a deeply ingrained emotional and psychological dependency—not a bad habit. It isn't going to let you go so easily.

This book teaches you the powerful mental skills of self-defense you need to fight and win this battle. It will help you regain your health and the freedom to shape your own destiny.

To win this life-or-death struggle, you have to know exactly who or what your opponent is. Something is attacking you, holding you down, taking away your health and your life. You are under assault—and you feel the affects of it every day.

But what is it? What is it that keeps you smoking, day after day, year in and year out, despite all your worry and pain and fear?

Is it the cigarette companies? All those clever and manipulating ads? Is it your own body? Or your hands, desperate to keep themselves busy?

Are *you* the enemy? Should you start beating yourself up, attacking your self-esteem by telling yourself, "I must be weak. I must be stupid—since it's so stupid to smoke. What's wrong with me?"

Cigarette ads, stress or low self-esteem can all trigger the desire to smoke, but they're not the crux of the problem. Your real adversary is an addiction to a drug—nicotine—and the cunning, manipulating voice *behind* that addiction. It is a voice that has a strange power over your mind, and it says things like this:

"You need me."

"You'll go crazy without your cigarettes."

"I'm the only friend you've got."

It is the voice that describes your future as an ex-smoker as bleak and lonely and joyless. It convinces you that the discomfort you experience in withdrawal will never go away, and tricks you with the promise that it will leave you alone if you will only give it "just one."

Nagging and relentless, unwilling to let you go, it seems to have a life of its own, stronger even than your desire for health—or your desire to live.

But as the years go by, another part of your mind gains strength and clarity. Desperate to protect you, a voice of intelligence and true self-love is beginning to plead, "You've got to stop smoking. It's frightening and painful. It's killing you."

Clearly, the time has come for confrontation, for standing up to the addiction.

But how, exactly? How do you overcome the deeply entrenched conviction that you can't live your life without cigarettes. After you stop smoking, what will you do when the craving to smoke comes over you—as it inevitably will? Just not think about it? Run away from it? Slap a nicotine patch over it, hoping for the best?

Your enemy has one agenda: to keep you smoking, to keep the addiction fed. What are you going to do: throw it a cookie?

Before going into this battle, you'd better be well-prepared. You'd better be well-armed.

This book confronts the problem of smoking at its core: the insidious grip that addiction has over your mind, the power it has to lure you back to smoking despite your best intentions and firmest resolve.

Here you will acquire an arsenal of self-defense skills, so that in any situation you are ready to fight off and disarm the craving to smoke, rather than be overwhelmed and defeated by it. You will learn that every desire to smoke is an opportunity to re-train your mind, so that your thinking works *for* you rather than against you.

CHANGING YOUR THINKING

For example, if you find yourself thinking "how great it would be to have a cigarette," you'd better not let the thought stand unchallenged. After all, how can a person be expected to feel good about *not* doing something if doing it would be so "great"? Right then and there you need to stop and take note of a different and more important fact: how *great it is* to be free, how *great it is* not to be coughing and worrying and hating yourself.

Or, if you're out with friends who still smoke, you may find yourself thinking, "They get to smoke, but I can't." Before you sink into a well of self-pity, let's look at the facts. Anyone can smoke—many thousands of people do, and most of them wish they could stop. So it's not that they *get* to smoke, they *have* to—all day, every day—and suffer the consequences. You don't have to. You're free.

For an ex-smoker, thinking how wonderful it would be to smoke is an illusion. If smoking was so wonderful, why did you stop in the first place? Why did you agonize about stopping for so many years before you finally got around to it?

Like letting go of any other addiction, stopping smoking involves a process of recovery. Essential to that process is being honest about what smoking was really like, as well as focusing on the benefits of being free from the addiction. Without a radical shift in your thinking, you won't have answers ready for the questions that can strike in moments of disappointment or depression: "Why bother? What difference does it make, anyway? Who cares?"

Such thinking can lead to relapse, so it's critical not to be caught off guard. There's too much at stake. In this book, I emphasize how important it is to be prepared for high risk situations and emotions, to train yourself to answer questions like these well in advance.

Why bother? "Because I don't want to be *bothered* by embarrassment, worry and guilt." What difference *does* it make? "It makes a difference in the way I breathe, the way my heart beats, the way I feel about myself. It makes a tremendous difference in every aspect of my physical and emotional health." Who cares? "*I care.* I care about my health, my freedom, my recovery. That's why I stopped smoking in the first place."

This new thinking may not ring true at first or vanquish all your doubts right away. But with consistent work, you take the power of words like "great" away from the addiction and give it to your recovery—your health, self-respect and well-being. Honest and clear thinking—just like any other habit—becomes normal, if you work at it.

If you're like most smokers, the greatest fear you have when thinking about stopping is that you'll *always* miss your cigarettes and that there will always be an emptiness inside you. You'll have lost your best friend.

But if you learn to face your addiction directly, challenging your old thinking, you'll soon begin to notice something. You no longer glamorize smoking or envy smokers. You've gotten over your love affair with cigarettes. You automatically look for the benefit of not smoking, rather than allow misleading thoughts—like "Wouldn't it be nice to have a cigarette"—to lure you back to smoking. You have literally "kicked" this monster out of your life.

Eventually—and it will happen much sooner than you expect—something astonishing is going to occur, something you thought would never be remotely possible. First a day, then an entire week, then whole months will go by, and you won't have thought about your cigarettes *at all*.

ABOUT THE CLIENTS PORTRAYED IN THIS BOOK

The clients who appear in this book are people I've worked with over the years. Their names and some details of their character have been changed to protect their privacy.

The excerpts of remember letters and thank you notes in the book were written by people who have taken my seminar in Portland, Oregon. Their names have been changed to protect their privacy.

HOOKED BUT NOT HELPLESS

Kicking Nicotine Addiction

Patricia Allison
With Jack Yost

PUBLISHER'S NOTE

CHAPTER

1

FIRE
AT MY BACK

A Helpless Smoker
Makes the Leap to Freedom

"Oh, it was agony."

"I was just climbing the walls."

"I stopped for ten days—I thought I was going crazy."

To hear people talk about trying to stop smoking, you'd think they'd been tortured on the rack. Or had their fingernails pulled out one by one.

Stopping smoking isn't easy. It's usually hard work. But it doesn't have to be torture.

If you stop smoking and go through pure hell, it's a hell of your own making.

I learned this lesson the hard way.

I smoked cigarettes for almost 20 years, and I was convinced I couldn't live without them. At the same time I knew I wasn't going to live very long *with* them, either.

I loved smoking, but I hated what cigarettes were doing to me.

I had a constant cough and a nagging pain in my chest. At times I'd be terrified when all of a sudden my heart would leap wildly out of rhythm.

Though I tried to stop many times, I always went back. I didn't understand why I kept doing something that I desperately wanted to stop doing. Was I crazy? Did I have a secret death wish? What was this love-hate relationship all about?

One morning I dragged myself out of bed, and as I lit my first cigarette I had a frightening realization: all of my attempts to stop had failed. I was still smoking. I was getting sicker and older, and smoking more. I was losing the battle.

Realizing what losing meant—being disabled, in pain and facing an early death—I panicked. I pushed all of my other preoccupations aside. Until I got a handle on this problem, nothing else mattered.

I decided to get help.

Clutching my cigarettes, I went in to a stop-smoking program, sure that only a miracle could save me. I had no idea what to expect. Would there be exhibits of blackened lungs? Or a film so frightening it would scare the devil out of even the most dedicated smoker?

The instructor assured us right away that he wouldn't waste our time convincing us that we *should* stop smoking. Obviously we already knew that or we wouldn't have shown up for class in the first place.

We wouldn't be getting any magic cures, either. We had an important battle ahead of us—a battle for life and freedom. And just as we'd failed when we tried to stop smoking before, we could lose this time, too. But now we'd have a fighting chance because we would be trained to meet the challenge.

The key was to get our thinking working for us instead of against us, to stop defeating ourselves, to avoid making the same foolish mistakes again and again.

We would need to clearly understand the nature of the problem. Chronic cigarette smoking was a true addiction—a drug dependency. We needed to understand exactly how nicotine worked and how we became so attached to it.

Also, a powerful industry aimed to keep us attached. We had to know how to fend off the barrage of seductive images that did their work so insidiously and so well, he said.

Defending ourselves against messages from the outside was hard enough. What could we do to defend ourselves against the tricks our own minds were going to play—sometimes months or even years later—to get us back to smoking?

After all, he said, anybody could stop, right? The problem was staying stopped.

This was a convincing, no nonsense approach. I was delighted, and it obviously showed. Half way through the first class, the instructor asked me why I was smiling.

"I just love what I'm hearing," I said quickly.

You bet I was grinning from ear to ear. I felt like a person who'd been teetering on the ledge of a burning building with fire at my back. And suddenly a safety net had appeared below.

I was smiling because someone finally had explained what the pieces of the puzzle were and how they fit together. What had seemed so hopelessly confusing and frustrating was beginning to make sense. My problem had a name, a pattern, characteristics—and a step-by-step solution. I was perfectly capable of controlling the urge to smoke, only I hadn't learned how yet.

I'd been hooked, but I was no longer helpless.

My enthusiasm and confidence grew with each of the five daily seminars. I was beginning to understand that the adversary was myself, the part of me that craved smoking. But another part of me wanted to survive, to live without pain and fear, to be healthy and free. And with the help of this program, I would finally be able to choose what I wanted most.

It was a struggle to take control of the compelling urge to smoke. It required effort and concentration. But it wasn't agony or torture.

In fact, I felt good about it right from the beginning. Rather than being crushed by self pity and fear of defeat, I was exhilarated by the power I now had to make a new choice for myself.

Winning is exciting. And for me winning this one was a turning point in my life. From then on, I knew the kind of work I wanted to do. I was fascinated by the impact that changing our thinking can have in all aspects of our lives. If I could bring other people the relief of finally having a choice, what a rewarding opportunity that would be.

That was over 20 years ago.

Since then I've taught thousands of people as desperate as I once was to stop smoking. While closely following the philosophy that I had learned, I took the program one step further. Understanding the exact nature of addiction and how to overcome it sometimes isn't enough. People may know exactly what they should do and still not be able to get themselves to do it.

Why? Because, I realized, they're blocked by a number of emotional and pychological barriers. Over time, I developed techniques to help people uncover their hidden motivations, fears and illusions. These mental traps, once brought to light, begin to lose their power. People feel liberated, more at ease, finally able to act in their own best interest.

This new strategy proved a powerful tool. More and more people were able to stop smoking, as my program expanded over the years.

You'll be meeting some of them in this book. They represent a typical stop-smoking class, a diverse group of people brought together by one common desire: to break their compulsion to nicotine once and for all.

When they arrive on Monday for the first of the five evening seminars, they take their seats with the wariness of people who are still strangers. As I look over the group, I reflect for a moment on how quickly a special bond will form between them.

Vera, a neatly dressed woman whose finely crinkled face betrays a lifetime of smoking, fidgets nervously to my left. Next to her, Bill, a young attorney who is tanned and alert, loosens his tie and sits back casually.

To my right is Jackie, a slightly overweight woman with a harried look, who has rushed into the room at the last minute and is still filling out her forms. Behind them, Stephanie, an attractive and confident newspaper reporter, is asking Lisa why she's decided to stop smoking at such a young age. At 22, with short spikey hair and feathery earrings, Lisa doesn't look like someone who would be worried about smoking quite yet.

I'm delighted to see Ellen again, quiet and self-assured, sitting next to them. A middle-aged, highly regarded counselor with her own clinic, she's repeating the class after a relapse. We had a number of fascinating conversations the last time she stopped, and she seemed to be doing just fine. But one day when I called her on the phone I was surprised to hear she'd gone back to smoking.

"I guess I just wasn't ready," was all she would say.

Lined up in the rear as far back as they can get are three men. Mike, a 42-year-old truck driver, sits with his arms crossed, with a look on his face that says, "What can this lady tell me that I don't already know?"

Norman, a Vietnam vet, is staring out the window as if he would rather be anywhere but here. Like the others in the class, he pretends not to notice the raspy breathing of George, the silver-haired man next to him.

As in most of my clases, some of these smokers are the hardest of the hard core. They're deeply attached to cigarettes; they've been using them as an emotional buffer against the stress and tension of life since they were kids. Yet even they will discover they can stop.

When clients have completed my classes, they often say, "This is more than just not smoking, isn't it?"

They're right. Stopping smoking correctly is about the power to take control of our own thoughts and feelings, rather than be torn apart by them. It's about learning to face reality, to take the difficult but necessary steps that healthy living demands of us. And it's about ending an obsessive love-hate relationship with a drug that in subtle ways stunts our emotional growth, cuts us off from our deepest feelings, and comes between us and those we love.

Above all, it's about winning self-respect, freedom, and the power to make new choices for ourselves.

"I can't smoke in the newsroom, anymore," Stephanie says. "How am I supposed to meet my deadline?"

CHAPTER

2

TRUTH OR CONSEQUENCES

Denial & Deprivation: Two Lies That Keep You Trapped

Have you ever wished that somebody would just lock you up somewhere so you couldn't smoke, no matter how much you wanted to?

If you're like most smokers, your answer will be, "Oh, yes, that's what I need!"—because deep inside you're not sure you can resist the temptation on your own.

But since nobody's going to imprison you, you do the next best thing. You put yourself in an imaginary jail and throw away the keys. You tell yourself, "Okay, that's it. I quit! I will never smoke again."

But even as the door clangs shut, doubts begin to creep in— and little by little they turn into panic. "Never smoke again? How am I even going to make it through the day tomorrow? I've got all those phone calls at work—and then there's that dinner party. Oh, no, what have I done? I've never been able to do it before, what makes me think I can do it now?"

In spite of your good intentions, you've set yourself up for a fall. Now that you've locked yourself up and quit forever, you *can't* smoke anymore, can you?

And if you can't have something you want—especially if it's something you think you desperately need—you're going to feel very unhappy about it.

DEPRIVATION

Almost everybody who tries to stop smoking walks into this trap. You jack up your willpower and throw away your cigarettes. Resolving never to smoke again, you sentence yourself to life without your "best friend." Cut off from your cigarettes, you inevitably begin to feel deprived.

Deprivation makes you feel miserable—left out, resentful, depressed, angry, full of self-pity and a sense of loss. And what do you do when you have feelings like this? You smoke!

You're caught in a vicious cycle. Here you are, trying to stop smoking, and the method you use creates feelings that make you want to smoke. To break this cycle, you need to re-examine your thinking.

What does it *really mean t*o be deprived?

It means that the thing you want or need has been forcibly taken away from you. Or that it's inaccessible, out of your reach. No matter how badly you want it, you *can't* have it.

But when you try to stop smoking, are you ever really deprived of cigarettes?

Not likely. Unless you're on a desert island, you can always get your hands on a cigarette. But as soon as you try to stop smoking, you find yourself feeling and acting as if you really were deprived.

What is the *thinking* behind this feeling of deprivation?

When you stop smoking and you see someone at the next table having a cigarette, what's the first thought that comes into your mind? "Oh, she gets to smoke, but poor me, I *can't*."

But that's not true. If you decide to stop smoking, you can always change your mind and start again. Yet the difference between self-denial and true deprivation is very difficult for people to grasp.

I CAN'T SMOKE

The first night of my class I ask my students, "When you've tried to stop smoking, tell me about a time when you couldn't smoke—when you were deprived."

"I went to a party and people around me were smoking," says Lisa, who in her punkish outfit looks like she's on her way to another party. "I felt left out because I couldn't."

"Oh, why couldn't you? Wouldn't anybody give you a cigarette?"

"Well, sure. But I quit the day before."

"You mean you couldn't smoke because you'd quit smoking?"

"That's right."

"Aren't you smoking now?"

"Yes."

"Wait a minute. I thought you said that because you'd quit smoking you couldn't smoke. If you can smoke now, could you have smoked at the party?"

"Well, I could have, but I felt like I couldn't because I'd quit."

"Aha, but you could have, couldn't you?"

She hesitates a moment before answering, "Yes."

"That's right, because you did later. You can quit smoking and still smoke—you can always change your mind."

Lisa was describing a choice she had made for herself, but acting as if she were deprived—as if she had no choice. Her actions proved otherwise.

"I'm deprived at work," Stephanie says with a touch of

resentment. "They won't let me smoke in the newsroom anymore."

"What do you mean? Does your editor confiscate your cigarettes at the door?"

"No—not yet anyway."

"Okay. Do you have cigarettes at your desk?

"Yes."

"Do you have a lighter?"

"Yes."

"Then you can smoke."

"No I can't."

"You mean you wouldn't know how to take out a cigarette and light it up, Stephanie?"

I look around the room. "Could somebody please show this woman . . ."

"Okay, okay," she interrupts. "I *could* smoke, but I'd get fired."

"So it's not that you *can't*, is it? It's that you don't because of the consequences. You're choosing to obey the rules in order to keep your job. Things change, don't they? Rules change. Let me ask you something. Can you spit on the floor at work?"

"Well, it wouldn't exactly enhance my image."

"I understand that. But can you, Stephanie? If you should have a sudden desire to spit, why don't you just go ahead and do it? Are there signs on the walls that say, 'No Spitting Here'?

"Well," I continue, amid the laughter, "at one time a lot of places had sawdust on the floor because spitting was very common. Then they started putting up signs in places where it wasn't allowed. And I can just hear the reaction, 'Where's our personal freedom going? You can't even spit where you want to anymore'.

"Times change, Stephanie. If you don't like the new rules at work, you can go somewhere else, can't you? Go work in a tavern. They don't mind if you smoke there."

Society can keep putting more pressure on you to stop smoking. But nobody can take your cigarettes away from you.

"What if they make tobacco illegal?" Mike says, from the back of the room.

"Then you could grow your own, couldn't you? Or smuggle it in from somewhere?"

Nobody argues with this point.

Vera raises her hand. She *knows* she can't smoke. Her doctor ordered her to stop after she had open heart surgery.

"When did you have your surgery, Vera?" I ask.

"About three years ago."

"And since then, have you been smoking?"

"Yes. About a pack and a half a day."

"Hold on a minute. I thought you said your doctor told you that you *can't* smoke because of your heart surgery."

Vera looks sheepish. "Well, I shouldn't smoke."

"Of course, you *shouldn't* smoke. But that doesn't mean you can't. What's your doctor going to do? Chase you around with a stick and hit you over the head every time you light up a cigarette? Nobody can stop you. You can smoke."

"But it might kill me."

"Sure it might. But you don't have to take care of your heart, do you? You can smoke instead. You've been proving that every day for the last three years. And you know something else, Vera? If you don't find a way to stop, you're going to *have* to go on smoking in spite of your heart."

Bill interrupts with an attorney's self-assurance. "I deprive *myself* of cigarettes," he says.

"And how do you do that?" I ask.

"I just throw them away and tell myself I won't smoke."

"How is that deprivation? Can you buy more and smoke if you want to?"

"No."

"Why not?"

"Because I won't let myself."

"Have you tried this method before?"

"Yes."

"Did you end up letting yourself smoke, anyway?"

"Eventually."

"You changed your mind, didn't you? You can stop smoking and let yourself smoke again any time you want to. That's not deprivation, Bill. That's short-term self-denial. Deprivation means you don't have a choice in the matter—you *can't* change your mind."

Consider the difference between a monk and a criminal. They both live in a cell, sleep on a mat and eat little more than bread and water. But, while the holy man is serene, the outlaw is miserable. One is making a free choice; the other's choice has been taken away.

Unless somebody locks you up and takes your cigarettes away from you, you're not deprived. For most of us, the twenty-four-hour convenience store is just around the corner. If you tell yourself you *can't* smoke, you're telling yourself a lie.

"So Bill, how could you truly *deprive yourself* of cigarettes? Here's one way. You're out in a rubber raft in the middle of the ocean. And you take out your last pack of cigarettes, tie a rock to it, drop it overboard and watch it sink. Now. Can you smoke?"

"No."

"Are you deprived?"

"Yes."

"Who deprived you?"

"I did."

"Can you change your mind now."

"No."

"That's right—and that's what it would take to deprive yourself of cigarettes."

It is essential to grasp this reality before you stop smoking. If you don't, you will inevitably fall into the torture of deprivation. You'll think, "I *can't* smoke anymore. I'll never be able to smoke again. What a horrible thought." And you'll begin to feel sadness and self-pity, a sense of loss. And thinking you *can't* smoke anymore leads to feelings of being coerced, of being forced to do something.

Then you'll get angry and start looking for someone to blame. "If I've *got* to quit, *somebody* must be responsible for this."

There are plenty of scapegoats. Your husband or wife. Your kids—haven't the schools turned them against you? A society that's been brainwashed by the Surgeon General. In fact, isn't everybody trying to take your cigarettes away from you?

By now you've totally forgotten your reasons for wanting to stop: the chest pain, the cough, the fear. Your anger builds on the feelings of loss and grief and finally explodes in rebellion. "Up yours, American Cancer Society! Nobody's going to tell me what to do."

Now your goal has become getting back what's been taken away from you. You're on a detour. You started out worrying about losing your ability to breathe. Now you're obsessed with losing your right to smoke.

Once you take the first step on the path of believing you are deprived, a natural progression of feelings lead you straight to relapse. In other words, you start off thinking you can't smoke anymore and end up smoking in order to prove you *can* smoke. What an accomplishment.

So when you try to stop smoking, tell yourself the truth. You can smoke. You don't have to stop smoking. And if you do stop, you can change your mind and smoke anytime you choose.

You don't have to give up your cigarettes. Of course, you have to give up something. It can be your peace of mind, your ability to breathe, and sooner or later, even your life. But it doesn't have to be your cigarettes.

"What are some other things you give up?" I ask.

"Sports," Mike says.

"Being able to walk up the stairs," George says.

"Good circulation," Stephanie says.

"That's right. And there are plenty of people out there giving up those things," I say. "Now I want you to start using this information. In the next two days before you stop smoking, every time you take a cigarette out of the pack, tell yourself, 'I can smoke. I don't have to stop and even if I do stop, I can start again'."

Norman speaks up for the first time from his corner in the back. "Hey, aren't you telling me to give myself permission to smoke. How's that going to work?"

"You've been giving yourself permission to smoke for years. That's what you're doing every time you light a cigarette. You've tried telling yourself you can't smoke, haven't you? How's that worked for you, Norman?"

"I guess that doesn't work either."

"Of course not, because it isn't true. What I want you to do is tell yourself the truth every time you light up. That way you cut down on feelings of panic and loss while you're getting ready to stop. And by reminding yourself 'I can always smoke,' you bypass all those nasty feelings of deprivation when you do stop."

Awareness is the first step in ending your relationship with nicotine. As a smoker, you've programmed yourself not *to think* when you light up a cigarette. You do it like a robot, like an automaton. You need to learn to think *before* you act.

THE BASIC CONFLICT

When you stop smoking, you'll have to face a conflict in yourself.

You desire two things: you want to smoke *and* you want to stop smoking. You crave cigarettes because you're addicted to them, and you want to stop using them because they're killing you.

Like most smokers, you probably try to solve this conflict by attempting to convince yourself that you don't really want to smoke. Since smoking is hurting you, you try to suppress your desire for it. In truth, it's not the *smoking* you don't want but the consequences of smoking: the cough, the pain, the stink, the fear.

It's very difficult to suppress a craving, especially for something being flaunted on billboards and indulged in by people all around you.

Yet you try. You attempt to stop smoking, and whenever the desire for a cigarette comes to mind, you push it away: "No, I don't want to smoke."

Since a compelling desire isn't banished quite so easily, you probably try to distract yourself and reach for a substitute. You keep busy, which usually means you keep busy *eating*.

But no matter how much you eat—or distract yourself some other way—the craving to smoke keeps coming back. The longer you deny it, the more tension and frustration builds. At some point, the truth finally bursts through: "I want a cigarette!" And you break down and give in.

At the end of the first class I challenge this way of thinking by writing a phrase on the blackboard: "I don't want to smoke."

"Is this true for you?" I ask each one.

Even if they smoke two packs a day and would kill for a cigarette, they usually say yes.

"And how long have you been telling yourself that?" I ask George, who at 65 has first-stage emphysema.

"Oh, about ten years."

"How's it working for you?"

"Not very well, I guess," he says, with a good-natured glance at the pack of cigarettes he keeps in the shirt pocket right over his heart.

"Of course it doesn't work. Because that's another lie. Look, what if they came out with a new brand of cigarettes called Mt. Breeze. Each cigarette is packed full of fresh air, vitamins and minerals. And when you smoke it you get a shot of health and vitality. Do you think you'd be trying to stop then? In fact, wouldn't your doctor be after you to smoke more?

"The fact is you want to smoke. That's what makes you a smoker. If you didn't want to, you wouldn't have a problem. You want to smoke *and* keep your health. But you can't have both. You want two things at the same time, and trying to deny your desire for *one* of them won't solve the conflict."

"But I don't want to smoke," George interrupts. "I hate it. It's just that I can't resist it."

"Oh, really? Well, let's see now. Name something else you hate—some kind of food."

"Squid."

"Good choice. Now, if there was a plate of squid in front of you, would you have to force yourself not to eat it? Would you have to take a special class to learn how to restrain yourself?"

"You've got a point."

"Right. Since you don't want squid, it's not irresistible— unlike smoking. You can hate the cough, the pain, the fear, but it's not the smoking you hate, is it, George? In fact, you *like* smoking, don't you?"

"Yeah, I enjoy smoking," he admits, to the amusement of everyone. "I just wish it wasn't killing me."

"Exactly. Now, there's something else I want you to do for tomorrow in addition to telling yourself, 'I can smoke.' Every time you light up a cigarette, tell yourself, 'I'm wanting to smoke right now.' Because that's the reality. That's what's happening to you, whether you like it or not. If you didn't want to smoke, you wouldn't be reaching for a cigarette."

WORDS ARE POWERFUL

You need to learn how to examine and correct your own thinking. Thinking creates feelings, which in turn influence your actions.

Words are powerful. They shape the way we think and feel about ourselves.

The surest path to relapse is to start from a false premise. Don't lie to yourself, telling yourself you *can't* smoke when you obviously can. That only leads to deprivation. You'll be miserable and end up smoking to escape the feelings of resentment, self-pity, and loss that you've brought on yourself.

And thinking "I don't want to smoke," is just as likely to lead to disaster.

To repress a craving takes a lot of effort—and repression only creates anxiety and stress. Under tension, the craving will spring out and overwhelm you. Instead of attempting to deny that you have a desire to smoke, acknowledge it and learn how to disarm it.

Be honest and clear about what's happening when you stop smoking: you have a choice. You can either give up your health or give up your cigarettes.

The truth is a powerful force in helping you make a choice you can live with. It will help set you free.

Bill stopped smoking on his own for six months, but at his high school reunion, he suddenly decided it would be nice to have "just one."

CHAPTER

3

THE
DRIVING FORCE

Nicotine: Fuel
for Compulsive Desire

"If you know smoking is killing you, why do you go on doing it?" I challenge my class. "Why do you want it so badly? Is it just habit? Something you do to keep busy? Something to do with your hands?"

"I think that's why I smoke," Vera says. "When I'm antsy or bored, I like to have something to do."

"Okay. Tell me the most desperate thing you've ever done to get your hands on a cigarette."

"Well, this doesn't sound too nice, but I've gone through the garbage before."

"Oh, I see. So there you are, you've run out of cigarettes at midnight, and you're down on your hands and knees clawing through the garbage, looking for a stale cigarette butt. What's that all about, Vera? Just keeping busy?"

As the others laugh, Vera also smiles at the absurdity of it.

"That's not why you smoke, is it—something to do with your hands? Well," I suggest, "maybe it's the glamour people want. Or that macho image." This time no one jumps at the bait.

"Okay, I know what it is. It's the taste, isn't it? Let's see. George, how many cigarettes do you smoke a day?"

"Oh, about two packs."

"Out of those 40 cigarettes a day, how many times do you stop and say to yourself, 'Umm, there it is, right there on the tip of my tongue, that down-home, Durham-North-Carolina-toasted tobacco flavor?"

"Are you kidding? None of them do that for me," George says. "The closest would be the one I have in the morning with my cup of coffee—and then it's only the first few puffs."

In fact, many smokers say that they hate the taste of cigarettes and often drink something, chew gum, or use mints to get the awful taste out of their mouth.

Obviously, it's not the taste or the need to fiddle with something that drives the compulsion.

Then what does compel people to go on smoking in spite of worry and self-disgust? What's behind this irresistible wanting?

UNWELCOME FACTS

Nicotine is a stimulant. It falls into the same pharmacological family as cocaine and amphetamines. In fact, in one study, when people who were familiar with the effects of all kinds of drugs were given injections of nicotine, most of them said, "Oh, I know that one, doc. That's cocaine."

In a 1988 report, former U.S. Surgeon General C. Everett Koop said, "Our nation has mobilized enormous resources to wage a war on drugs. We should also give priority to the one addiction—tobacco addiction—that is killing more than 300,000 Americans each year."

Most people assume that the "hard" drugs are more addictive than cigarettes. But two experts, Dr. Jack Henningfield from the National Institute of Drug Abuse and Dr. Neal

Benowitz from the University of California at San Francisco, in separate 1994 studies, compared nicotine to heroin, cocaine, and marijuana. Not surprisingly, both experts ranked nicotine as the least intoxicating of the four drugs—but the *most addictive* of them all.

Like those addicted to heroin or cocaine, nicotine addicts use their drug compulsively, develop tolerance to it and suffer withdrawal symptoms when they try to stop. Although many smokers quit on their own, they also show relapse rates similar to those seen in addicts who withdraw from illegal drugs.

So that's what's behind the wanting: addiction. That's what's going on when you're digging through the garbage at midnight, when you go out in an ice storm and risk life and limb to make it to the comer store—you're after your drug.

One of the reasons we get so attached to nicotine is that it's biphasic—it has two phases. It acts both as a stimulant and as a depressant, creating a state of relaxed alertness. That's why you can use smoking both to calm yourself down or to pick yourself up, which makes nicotine a pretty handy little substance.

Unlike speed or heroin, nicotine takes you neither too high nor too low to function normally. The drug effect is just enough to take the edge off stress or boost your spirits at a party.

And you've learned unconsciously how to give yourself just the right dose. When you're rushing around trying to get things done, you take short little blasts to give yourself a lift. But after a hard day's work when you sit back in your favorite chair you take long deep drags to relax.

In other words, you're using nicotine for the same reasons other addicts use their drugs—to control your moods. When you get angry or anxious, you calm yourself down with a few shots of nicotine and you feel better. You're using it as a tranquilizer.

Though most people have heard that smoking is an addiction, the implications have never really sunk in.

"The truth is," I tell my clients, "if you have to smoke ten, 20, or 40 cigarettes a day in spite of pain and fear, you're a drug addict. You've got a monkey on your back.

"Every time you feel that urge and reach for your pack of cigarettes, you take out one of those little white 'disposable syringes' and inject your drug. In a few seconds you're loaded. Now you can get back to your life because you've got your fix.

"You're helpless to resist the urge to smoke. You can agonize and hate yourself and vow never to smoke again—and you'll take that next cigarette anyway. You have to or you'll go into drug withdrawal. Just like any other junkie."

This is tough talk, but it's necessary to get people to face the seriousness of their predicament. Otherwise they'll go on treating smoking as a naughty little vice or a bad habit, underestimating it and therefore undertreating it until it's too late.

Forced to look at their behavior in such graphic terms, many of my clients are disturbed, shocked, angry—and sometimes even brought to tears.

And why shouldn't this news be shocking? Cigarettes have never been marketed or sold as a drug. Warnings on cigarette packages have only included the danger of becoming addicted in recent years.

What's more, the effects of nicotine are very subtle. You've been using it for so long you've built up a tolerance to it. And the nicotine is dosed out in such small amounts, you're at maintenance level all the time.

When you first started, you might have felt mildly euphoric, dizzy or excited. But these were only funny side effects; you sure didn't think of this as "getting high."

"It is far easier to become dependent on cigarettes than on alcohol or barbiturates," claims Dr. Hamilton Russell, a researcher at the London Institute of Psychiatry. "It requires no more than three or four casual cigarettes during adolescence virtually to ensure that a person will eventually become a regular dependent smoker."

Not everyone who smokes is addicted. Social smokers smoke

when they have a spontaneous desire. But such people are rare—only three percent of all smokers, according to some studies.

"One hallmark of an addicting substance is the fact that users seek it continuously, day after day," according to the Consumers Union Report, *Licit and Illicit Drugs*. "If they can take it or leave it—take it on some days and not be bothered by lack of it on other days—they are not in fact addicted. Judged by this standard, nicotine is clearly addicting; the number of smokers who fail to smoke every day is very small."

Every time you take a drag off a cigarette, you inhale the drug nicotine. It hits the brain within seven to ten seconds—twice as fast as intravenous drugs and three times faster than alcohol.

At that point, the craving for nicotine is relieved, and the relief can last anywhere from 20 minutes to an hour, depending on your tolerance.

Then the craving for nicotine returns and you become obsessed with getting that next fix. Without another cigarette, you become anxious, nervous, tense: symptoms of withdrawal from an addictive drug. It's a form of slavery.

When you smoked that first cigarette as a teenager, you were making a choice. For a time, you could take it or leave it. But at some point, the addiction began to tell *you* when to smoke. You no longer had anything to say about it.

Though you never intended it—never dreamed anything like that would happen—you began a lifetime drug dependency.

WHO'S TO BLAME?

Smoking isn't something you do because you don't care or because you're just too pig-headed to stop. You smoke because as a teenager you played around with cigarettes and you got hooked. You didn't know any better.

Once you realize what's running you, you can stop kicking yourself around because you haven't been able to put down your cigarettes and walk away.

You don't have to feel guilty or bad about yourself. If you'd known you were going to get addicted, you never would have started. You were an innocent kid, ignorant of the facts. Through no fault of your own, you invited a craving into your life. It has moved in. Now it's yours to deal with.

ADDICTION: PSYCHOLOGICAL, EMOTIONAL, PHYSICAL

Nicotine addiction is a psychological-emotional relationship with a drug. Smokers become dependent on tobacco as a mood and mind-altering substance. They come to love smoking, even though they hate what it's doing to them.

However, most people are used to thinking about addiction as something that is mainly physical, that their bodies are addicted and crave the drug. After all, they've almost all tried to stop—and felt the physical distress of withdrawal.

It would be a wonder if our bodies didn't react. Nicotine is a highly poisonous substance, commonly used in insecticides. The body has adapted to the use of this poison year after year and must re-adapt when you stop using the drug. That's the main cause of physical symptoms during withdrawal.

But your experience of withdrawal depends on how you approach it. If you stop smoking and you feel like you're going crazy or "climbing the walls," it has nothing to do with your body; you've got yourself mentally locked up. Coming off a drug you've been using for years is uncomfortable. But it's torture or agony only if you allow yourself to feel deprived.

This much is obvious: it's not physical addiction that makes you take that first cigarette months or years after you've stopped smoking. The unpleasant physical sensations—that occur for a few days during withdrawal—have been long over with.

However, the emotional and psychological addiction hangs on.

Just how deeply attached people are to cigarettes depends on many things: their childhood and personality, the age they started, how long they've been smoking, the demands of their daily lives, and the progress they've made toward resolving emotional problems.

People with a healthy upbringing who take up smoking as adults, for example, may find it fairly easy to break the attachment on their own. Stopping smoking for them is mostly a matter of readjustment, not an agonizing experience. Once free, they feel relieved, happy to be in control and improving their lives.

While stopping is easy for some and almost impossible for others, most of us fall somewhere in between. We have a tough time ending this love-hate relationship and need all the help we can get.

Smokers often think of cigarettes as their best friends. It's sad when Vera, a divorcee whose two children live in another state, tells me, "I feel like cigarettes are all I've got left." Many people feel the same way.

How did we become so attached to cigarettes? Why do we think of them as such good friends? Why do they seem to fill an emptiness inside us, to take the edge off a vague hunger and craving within?

We're attached because from a young age we've formed an emotional bond with cigarettes—with the stimulating and calming effects of the drug nicotine. Repeating this ritual tens of thousands of times over the years, we learn to depend on cigarettes to manage our emotions. They make us feel safe when we're anxious, less lonely when we're cut off from others. They soothe us when we're irritated or angry, occupy our time when we're bored, give us a shot of energy when we're tired.

Smokers aren't used to facing life—with all its emotional pains, frustrations, disappointments, tediums, even moments of elation—without the stimulating or soothing effects of nicotine.

"Tell me about taking that first cigarette when you relapsed," I ask Ellen. "What did you get out of it?"

"Not much, really," she says. "It sure didn't taste good."

"Then why did you keep on smoking? How did it make you feel?"

"It made me feel dizzy."

"Is that what you liked about it?"

"No, I didn't like that feeling. I'm not sure why I did it."

"How were you feeling at the time you smoked?"

"I guess I was anxious and upset. I was trying to sell my house and running into problems."

"Okay. What did smoking do for you?"

"It made me feel better. It calmed me down. I wasn't as nervous."

That's how nicotine works. It doesn't have to bring you any great pleasure; you may not even like the physical effects. But it's emotionally soothing. It takes the edge off unhappy feelings.

Bill mentions that after seeing his uncle die of throat cancer, he stopped smoking on his own and went for six months without any problem. He was very happy to see his health improve. He got his wind back, began swimming and working out, and was feeling great.

Then, at a high school reunion, he saw a couple of old friends who were smoking and decided, "That sure looks good. I think I'll try one."

It's evident that the urge wasn't his body demanding a hit of nicotine. It was the junkie in him, wanting to enhance his mood, wanting that "drug experience."

ONE PUFF

"And by the way, Ellen," I continue, "let me ask you something. When you took that first cigarette after three months, did you intend to go back to smoking 30 cigarettes a day?"

"It was the farthest thing from my mind."

"What did you intend to do?"

"I was just going to smoke one so I wouldn't feel so anxious."

"Then what happened?"

"It just made me want another one."

"What about you, Bill? Were you tired of all that swimming and better breathing and feeling good about yourself? Just couldn't stand having peace of mind and health and decided 'Hey, I'm going to give this up and be a smoker again?' Is that how it happened, Bill?"

"Oh, sure. I couldn't wait to feel run down and lousy ," he says, laughing.

"In fact, Bill, you didn't intend to lose those good feelings at all, did you? You were just going to have fun at the reunion and go on being an ex-smoker. Well, I've got news for everybody in this room," I say. "Once you stop smoking, your biggest mistake will be to try to act like social smokers, people who just smoke occasionally, when it strikes their fancy. George, how many cigarettes do you want each day."

"About 40. Sometimes more."

"Then isn't it incredible that you'd be able to convince yourself that you'd be satisfied with just one and not think about the other 39? If you only wanted one, why aren't you out there smoking one? You'd be kidding yourself. You want them all.

"How about you, Vera? Are you ever satisfied with one?"

"Never. The ladies I play bingo with at church ask me why I don't just smoke when I really want one. Well, I tried that, but I really want one all the time."

The class breaks up over this; everyone knows exactly what she means.

"That's because when you smoke, Vera, you smoke compulsively. Social smokers aren't hooked—at least, not yet. They don't have to try to get control of their smoking because it hasn't gotten out of control. But nicotine is your drug of preference. It works for you and, unfortunately, it's very addictive."

Once you're addicted, why does one puff inevitably lead back to full-time compulsive smoking even if you haven't smoked for a long time?

First, because over the years you've conditioned yourself—chemically—to tolerate a certain level of nicotine.

When you're anxious or depressed and use nicotine to feel better, your body reacts to protect itself against this highly poisonous substance. Ironically, while the body is busy trying to defend itself, you're getting high. The heart speeds up, adrenaline pours into the blood stream, and the brain releases certain pleasurable chemicals to serve as natural painkillers, so you feel more alert, more confident, and at once more relaxed and energized.

You have found a clever way to exploit the fight-or-flight response normally activated only by sudden dangers in the environment. But you are triggering this reaction 20 or 30 times a day in order to handle routine anxiety.

In an attempt to restore balance, your brain gradually develops a tolerance to the drug. It doesn't react quite so dramatically anymore. Eventually you reach a limit. If you smoke more you make yourself sick; if you smoke less you start getting edgy. You're stuck with an uneasy truce. You don't get the buzz that you used to, but you settle for what you do get: a mild relief from tension you can always count on.

Once you've stopped smoking for a while and your body has readjusted to its natural state, taking the first puff triggers all the initial reactions to nicotine. You get the kind of drug experience that is certain to make you want more. But with so many years of training, the body never forgets how to protect itself against nicotine. So it's only a short time before you're right back where you left off—smoking more and enjoying it less.

There is another reason one puff leads back to full-time smoking. Once you choose to handle tension by using a drug, you want to keep doing it.

That's what Ellen wanted: relief from stress. And she got it. Even though smoking didn't resolve her problems, it did make her feel better for the moment. And since she'd made it okay for herself to smoke one, what could be easier for her than to make it okay to have a few more?

And besides, smoking wasn't hurting her yet. She was getting the benefits of the drug without all the miserable consequences of being a smoker.

Like Ellen, it's easy to fool yourself if you think you can smoke "just one." When you haven't smoked for a while, the first one stands out as a positive experience. The next time you have a desire to smoke, you'll think of the instant gratification you got from smoking that one, rather than the misery of smoking 30 cigarettes a day. And you'll go for it again.

When you've only smoked a few cigarettes, you'll be lulled by the feeling that nothing bad is happening. But soon, every incident in your day seems to call for a cigarette, just as it always did. Before you know it, you're back in the grip of full-fledged addiction.

You start noticing those little nagging symptoms again, and you think, "I've just got to quit doing this real soon—but not right now."

I tell my class, "If you ever get yourself to that blessed place where you have stopped using your drug, there's something you need to know. One puff and you'll go back to smoking 20 or 30 cigarettes a day, every day, whether you like it or not. How do you know this is true, Ellen?"

"Because that's exactly what happened to me."

"That's right. And what about you, George? You've never stopped smoking for more than a few hours in 50 years, have you? How do you know this is true for you?"

"Well, I don't, but I guess I'll have to believe what you say."

"Believing is not good enough. You need to find a way to know that this is true." I ask the others, "How many of the rest of you know from experience that one puff will take you back?"

Everyone raises their hands.

"That's pretty good evidence right there, isn't it, George? But you need to figure this out for yourself. The law of compulsion is like the law of gravity. You don't have to jump off a ten-story building in order to prove you're going to fall, do you?"

There is nothing more important to grasp than this: the crux of beating a drug addiction is knowing it's the first fix that does you in.

FOREVER SUSCEPTIBLE

Getting through withdrawal is just the first phase of controlling any drug addiction. Withdrawal is uncomfortable, but only temporary.

Even late-stage alcoholics or heroin addicts can make it through severe withdrawal when they have enough motivation. The real problem of addiction comes with the second phase: staying free.

Once you've stopped smoking and gotten through withdrawal, you'll inevitably have desires to smoke in situations that you associate with smoking.

Even months or years later, with no nicotine in your body, you can still experience a sharp—though usually brief—urge to smoke, especially in an emotional crisis.

It's as if a trap door is at work. When enough time has passed, the desire to smoke sinks below consciousness. The door closes. In a crisis, it suddenly snaps open, and out springs the magical solution: "A cigarette would make all this better."

If you know what to expect and resist this urge, it soon goes latent again. But it's still there—just beneath the surface—a method of coping you never completely forget. It remains etched in the memory bank.

Once an addict, always an addict.

Like diamonds, but without the sparkle, addictions are forever.

Anyone seeing Lisa in the local coffee house, smoking and laughing with her friends, would be surprised at the 'remember' letter she wrote in class.

CHAPTER

4

WHEN YOU'RE READY

Method & Motivation: What Works & What's Worth Fighting For

"Do you mean because I'm addicted, I'm always going to have to worry about this thing?" Lisa asks, at the end of our discussion about nicotine. "It's never going to be over?"

"That's sounds awful, doesn't it?" I say, "Especially for someone as young as you. Is anybody else in here worried that for the rest of your life you're going to be wishing you could have a cigarette?"

Every hand goes up—except Ellen's.

"You may find it hard to believe," I say, "but that's just not going to happen, because you're going to learn a better way. Ellen, you've worked with this program before. What was your experience?"

"Well, I was certainly obsessed with smoking for the first few days—really dysfunctional. I even had to reschedule a couple of clients. But I kept working at it, and suddenly my desires to smoke dropped way off. After a few weeks, I was surprised to find that I was going for days without even thinking about it at all."

"And the best part is that if you hadn't taken the first cigarette," I say, "you'd soon have gotten to the point where you hardly ever thought about it."

"Then I'm not going to have desires to smoke for the rest of my life?" Lisa asks.

"You'll have them, and you'll need to deal with them when they come up. But they'll be less frequent and less intense as time goes on until you'll feel like a non-smoker. You won't have to worry about relapse as long as you give up your illusions about being able to smoke just one."

MOTIVATION

Human beings tend to resist change; and the more radical the change, the more they resist it. In order to stop smoking, you've got to be strongly motivated. It's not enough simply to find it a bother.

Since smoking is woven into every aspect of daily life, no wonder so few people can just put down their cigarettes and walk away. A lot of people would like to stop, but they're not willing— they're still getting away with it. They haven't "bottomed out yet," to borrow a phrase from Alcoholics Anonymous. Though uncomfortable, they're not experiencing enough real pain or fear.

If you're not scared enough or hurting enough, you might make a half-hearted effort, but you won't change. Compulsion has a powerful hold on you, and you've got to be willing to struggle, to do what you have to do in order to save your life. You're not going to get anywhere until you've reached that point.

Being adequately motivated means being clear about just what you're giving up and what you hope to gain.

Before you decide to stop, take the time to make a list. Write

down each reason you have to stop smoking. Be specific.
Which of the symptoms below are you experiencing already?

sore throat	obsession to stop
cough	embarrassment
phlegm	slavery
nausea	guilt
chest pains	self-hatred
heart problems	fear and worry
heart palpitations	low energy
indigestion	emphysema
shortness of breath	bronchitis
sinus problems	congestion
circulation problems	headaches

Simply noting these words may not have enough of an impact on you. To make it more real, write yourself a "remember letter," detailing what smoking was really like.

Describe for yourself not only the physical effects but also the emotional price you've had to pay: the worry and fear, the endless attempts to stop, the feelings of failure and self-loathing, guilt and despair. Try to capture specific moments when you were especially unhappy with smoking.

Most of my clients appear to be healthy individuals, content with their lives. But their remember letters, which they write during the course of the seminar, prove how deceptive looks can be.

My current class is no different, except for George, whose breathing is audible throughout the room. Anyone seeing Lisa in the local coffee house, smoking and laughing with her friends, would be surprised at the letter she wrote in class.

"Remember the time you went to the doctor," she writes, "and he told you that because of your asthma, smoking had caused permanent damage to your lungs *already*. Remember all those nights not being able to get to sleep—gasping for breath. Remember the fear that you might stop breathing altogether in your sleep!"

Norman, at age 40, writes about getting half of his right lung removed. Jackie remembers questioning whether she would live to see her daughter graduate from high school, and how guilty she feels about the cough that her husband said went on all night.

Stephanie recalls waking up at 4 a.m. nearly every morning, agonizing about how she was ever going to stop—and smoking while she agonized. "Remember how humiliated you felt being one of the last reporters in your office who was still smoking, the despair that followed every attempt to quit, the embarrassment of being a slave to cigarettes."

This side of the story would definitely be news to the people in her office who had so often heard Stephanie defend her right to smoke.

Bill looks sharp and in control of his life, but he wonders how long he's going to be able to keep up with his law partners when they play racquetball. And he reminds himself "about that raw, searing pain" in his chest some mornings when he takes a deep breath, and how he tries to delay his first cigarette because he knows the first few puffs are going to hurt.

And because Ellen is repeating the program, she writes a new letter: "How did you start smoking again? You weren't paying close enough attention. Maybe you didn't believe you were really addicted. Imagine that! But remember how horrible you felt when you realized you were smoking again—the feeling of utter despair that seemed to undermine everything else in your life, to cast a dark pall over your work, your relationships, all your success and accomplishments."

This simple exercise—which takes only about ten minutes—is extremely important. Think of it as part of your insurance policy for the future. After a few months, when you're feeling great and you start thinking smoking wasn't so bad, reading this letter will bring you back to your senses. You'll see exactly how bad it was and why you were obsessed with stopping for so many years.

BENEFITS

Once you've looked at what you don't want, you need to start focusing on what you *do* want for yourself. Choose from the list below ten major benefits that you hope to gain.

Better Breathing
No Shortness of Breath
No Gasping for Breath
No Chest Pain
No Wheeze
No Congestion
No Cough (or Hack)
No Phlegm
Arrest Lung Disease
No Bronchitis
No Hoarseness
Clearer Throat (Voice)
No Sore Throat
Lower Risk of Throat Cancer
Stronger/Healthier Heart
No Heart Palpitations
No Panic
Lower Risk of Heart Attack
Clearer Sinuses
Less Headaches
No Runny Nose
Better Circulation
Lower Risk of Stroke
Lower Cholesterol
Lower Blood Pressure
More Energy
Greater Endurance
No Drained Feeling
Be More Active
Better Sense of Taste & Smell
No More Cigarette Hangover
Awaken More Refreshed

More Motivation
Less Gum Disease
Nicer Breath, Cleaner Teeth
Better Immune System
Less Colds or Flu
Better Digestion
No Nausea
Better Complexion
Less Wrinkles
Better Night Vision
No Worry or Fear (of lung
    cancer, stroke or emphysema)
No Self Hatred or Disgust
No More Feeling Stupid
Sense of Control
Greater Self-Respect
No Embarrassment
Less Depression
Less Dirt (Ashes, Butts)
No Stink
No More Burn Holes
Laugh Without Coughing
Sing Again
Freedom to Sit Where I Want
No More Obsession to Stop
Get On With My Life
More Closeness (with Friends,
    Family and Lovers)
Better Sex and Intimacy
No Constant Conflict
Better Outlook on Future
Save Time and Money

Are the benefits you chose vitally important to you? Are they worth fighting for? Don't put minor benefits at the top of your list. Nobody stops smoking for clean ashtrays.

Be sure these are your reasons—not benefits you think you *should* want, or other people's reasons. Search within yourself for what you want. How does having to smoke affect you every day? What do you want for yourself from not having to smoke?

A good time to write down your reasons for stopping, your remember letter, and your list of benefits is the day before you stop smoking.

TIMING

What doesn't stop, goes on. If you wait for the perfect time to stop smoking, you'll never stop. Though no time is perfectly free of stress and demands, some times are better than others, and some are almost impossible.

Do not attempt to take on the job of kicking this drug addiction during an emotional upheaval—in the throes of losing your job, breaking up a relationship, coping with a crisis in the family—or with some disaster looming on the horizon. Under such conditions you can hardly expect to muster enough energy and concentration to stop smoking. And if you try to stop while taking a vacation, probably the only thing you'll accomplish is to ruin your vacation. An atmosphere of indulgence and relaxation hardly inspires self-restraint and rigorous effort.

I'm always amazed when people come in to my program in the middle of a painful divorce or custody battle or shortly after a loved one has died. The only reason they attempt this is because they think they only have to conquer a bad habit. They don't realize they are about to confront a powerful drug addiction.

SHOULD YOU PAY FOR HELP?

Often people hesitate to pay for help to stop smoking. Since they underestimate the problem in the first place, they're reluctant even to admit that they need help, let alone pay for it. On the contrary, most smokers are so attached to their smoking that they think somebody should *pay them* to give it up.

Since no program can guarantee success, people who don't stop smoking with a particular approach may feel as if they've wasted their money. It's true that some programs are *much* better than others, and it's important to try to get the best help available.

But any money you spend toward finding a way out of this terrible dilemma won't be wasted. You're going to learn something in any program—something that, as your knowledge and experience grows, will eventually help you to stop smoking. Even if you only discover what *doesn't* work, then you're one step closer to finding out what does. If you go to a hypnotist expecting a quick fix, maybe you'll learn that a quick fix won't work for you.

Don't be so afraid of spending money on this problem. Every time you light a cigarette, you're proving you've got money to burn—doing yourself in. And until you find a solution, you're going to have to go on burning it up whether you like it or not.

You may not want to pay for a stop smoking program, but if you're a typical smoker, you're going to have to pay somebody. Right now, you're paying the cigarette companies anywhere from 60 to 100 dollars a month. And what about the throat lozenges, special toothpaste, sinus medication, aspirin and nasal sprays? What about the extra cleaning bills and higher insurance rates? Remember the time you burned a hole in that brand new silk dress or down jacket?

That's what smoking is costing you now. Eventually, you'll be paying hospitals and surgeons.

Doesn't it make more sense to invest a modest sum now to save yourself thousands of dollars in the future? Money spent to stop smoking is an investment—one of the best you will ever make. Even if solving your drug problem costs you $500, you'll save more than that in one year alone. And over a lifetime, at today's prices, an average smoker will save as much as fifty times that amount in cigarettes and related costs. What other investment gives you that kind of return?

And if it gives you back your health, your peace of mind, and your freedom, it's the bargain of your life.

After stopping, Vera included this note with her payment for the class: "This is one check I actually enjoy writing, since it reminds me that I am finally free!"

AN EFFECTIVE METHOD

If you've tried to stop smoking and failed, you've probably been told you didn't want to stop badly enough.

I know this isn't always true. People come into my class who've been working intensely with the problem a long time. It's obvious that they really want to stop; they're obsessed with the idea.

Stephanie, who was once assigned a story about stop-smoking programs, not only wrote about them but tried them all. "I've tried every program that's come through town," she said. "The longest I was able to stop was two weeks."

Though many people have plenty of motivation, they don't have the right method. They've got the *will*, but they haven't yet found the *way*. When they do find the right approach, they do stop smoking.

An effective method begins by treating the problem for what it is: drug addiction—not simply a bad habit. To break this addiction, you need to learn skills for coping with the discomfort of withdrawal, sudden urges to smoke, crafty and self-defeating rationalizations, complacent overconfidence, and emotional crises. An effective approach will teach you the reality of drug dependency. While you can live comfortably and happily without smoking, you'll need vigilance to avoid relapse, even after many months or years.

That's why the best methods include long-term guidance and instruction. Based on your particular needs, this follow-up work helps correct dangerous thinking and reinforces the new skills.

If your program doesn't offer in-depth follow-up, you may find help through the group Smokers Anonymous.

MAKING THE BREAK

One advantage of taking a stop smoking program is that you have a definite time set aside when you are going to stop. In my own seminars, I ask my clients to stop in the middle of the week, before the Wednesday night session.

"Have you stopped smoking?" I ask each person at the beginning of this class, right after they've smoked their last cigarette in the hallway.

"I hope so," Vera says. "But I don't know if I'm going to make it."

"Wait a minute. I didn't ask you about the future. I only want to know, have you stopped smoking now?"

"I guess so."

"That's not good enough, Vera. You can always change your mind later, but right now you've either stopped or you haven't. Which one is it?"

"Yes, I've stopped."

"Great. Congratulations. We can work with that. You've got to stop before you can go on to the next step. So right now you can feel good about what you're doing, because you've already accomplished something very important, something that's going to make a tremendous difference in the quality of your life."

"What should I do with my cigarettes?" Jackie asks, holding up her pack of cigarettes.

"Good question. What do you normally do with your cigarettes when you try to stop smoking."

"I throw them away."

"How far away is far enough so that you'll be safe from them? You can't get away from cigarettes in this society, can you? And here's another problem. When you throw your cigarettes away, how do you feel when you find yourself somewhere without them? How about it, George?"

"I panic."

"That's right, because you've had your drugs right there in your shirt pocket for the last 50 years, haven't you? And if you suddenly throw them away and find yourself stuck in the middle of a traffic jam, you're going to go into instant deprivation. You're going to panic because you can't smoke. Your mind will be obsessed with one thing: getting your hands on a cigarette. You'll be unable to think about your *real* problem: how to work through that desire to smoke and stay off smoking."

"That happened to me once," Mike says. "I actually got out of my truck and went around to about five cars before I found a cigarette. My buddy couldn't believe it."

"I can believe it," I assure him. "I've heard a lot worse. So the best thing to do is to keep your cigarettes with you."

Vera looks alarmed. "Oh, no, I'll never be able to do that."

"Sure you will. You're scared because you still believe that if you're around cigarettes you have to smoke them. But that's not true. In fact, if you don't learn how to be around cigarettes without smoking them, you're doomed. Because they're everywhere. So, keep your cigarettes with you, at least for a few days; this will help you stay out of deprivation."

"Of course you don't *have* to keep them," I continue. "Get rid of them if that makes you feel more comfortable. But remind yourself that you can smoke and that the convenience store is only five minutes away. You're not deprived. If you're like most people, though, you'll come in tomorrow, saying, 'I feel fine as long as I know my cigarettes are right there.' But there is something else you can do with them that will help you. Everybody please take out your cigarettes."

When everyone has their cigarettes out, I say, "Look at those nice, neat little packages. So shiny, with those bright, cheerful colors and classy emblems. Mike, George, you guys always carry yours in your shirt pocket, don't you, so they're safe from harm. And Bill's come in a crush-proof box—so that none of those precious little syringes will get broken, right? Oh, and look over here. Stephanie's got a special leather pouch to keep her drugs in, with a side pocket for the lighter so it's always handy. Nothing's going to happen to your stash, is it, Stephanie?

"Now, Jackie, let me have your pack for a minute. And all of you, take your cigarettes in your hand like this, and give that nice neat package a good squeeze," I say, demonstrating. "Don't break them up, just make them look like garbage. Because that's what they are."

Some of the people squeeze their packs with a vengeance; the others grimace painfully.

"Now they don't look like those glamorous, seductive images in the ads anymore, do they?"

"But if I still have them with me," Jackie says, "I might light one up without thinking about it?"

"Take a look at this, Jackie," I say, returning her pack, which is now twisted into an unnatural, oblong shape. "When is the last time you carried around your cigarettes in this condition? Do you really think you could take out a cigarette without knowing you were doing it?"

"By the time I unwadded this mess," she says, laughing, "I guess I'd know what I was doing."

"You sure would. You're not a robot. And if you think about it, you'll realize that anytime you stopped smoking in the past, you always knew you were doing it when you took that first cigarette. Once you stop, you have to make a deliberate choice to smoke. You don't 'slip'."

MAKING THE EFFORT

People often resent the time and effort that it takes to successfully stop smoking. They think somehow the problem should just go away. They're too busy and have more important things to deal with in their lives.

But look at how much time and energy you devote every day to smoking—running out of cigarettes and running to the store for more, keeping track of cigarettes and cleaning up after them, worrying about smoking and planning to stop, taking constant breaks and looking for a place to smoke. And what about nursing a "three-day cold" that drags on for three weeks?

People who expect to stop smoking effortlessly become discouraged when they don't achieve instant success—especially if they see someone else stop without any apparent trouble. But everyone has a different degree of attachment to smoking, and some will need more help than others.

Like it or not, you're going to need to work with it until you find your way out. It's not that you can't stop, it's only that you haven't yet learned how. Until you do, don't resign yourself to being chained to cigarettes the rest of your life. The key to freedom is the desire to be free.

No matter how long it may take you, no matter how many setbacks or failures you may have—the rewards of not having to smoke are worth it. They're worth all the effort, expense and investment of time.

But to make that effort, you'll have to keep your goals clearly in mind. The things you want the most for yourself—your health, self-respect and peace of mind—must triumph over your wanting to smoke.

"Well, I'm not smoking, but I can't say I'm doing it willingly," Mike says.   "I'd call it *begrudgingly*."

CHAPTER

5

BREAKING
THE CONNECTION

The Temporary Discomfort of Withdrawal:
Taking Control vs. Going Crazy

Sometimes it seems like you just can't win. When you're smoking, you've got a problem—wishing you could stop. After you stop, you've got a new problem—wanting to smoke.

I ask my class what they do about the wanting when they stop on their own. Every one of them talks about trying to get rid of it.

Bill says he likes to jog; he even keeps running gear at his law firm, so he can get out during the day. "When I get a real craving to smoke, I just drop everything and go for a run."

"And after you get back, what do you want to do then?"

Bill hesitates. "It doesn't make sense—but I always want to smoke."

"That's right. You can't run away from this one, can you, Bill? Because that urge will be right there waiting for you when you get back."

Vera, to no one's surprise, talks about keeping busy. "I clean out the closets and wash the windows and polish the furniture—I just keep going."

"Great, but you get tired, don't you? What happens when you

sit down to take a break?"

"I don't take a break."

"Vera, you can't work non-stop for days. You're going to get tired. And as soon as you step back to admire that polished furniture, what's the first thing that's going to come to mind?"

"Oh no," she says, getting a picture of something she's done countless times. "I'm going to want to smoke."

"You sure are. None of this stuff is going to work, is it? You finish cleaning your house, and you want to smoke. You get bored watching TV, and you want to smoke. After a snack, you want to smoke. These things don't get rid of the desire to smoke, they *call* for a cigarette."

TEMPORARY DISCOMFORT

People try to get rid of the desire to smoke because they don't know what else to do about it. It makes them uncomfortable. But the answer is simple: Don't try to get rid of the discomfort. Accept it.

My class reacts to this revelation suspiciously. What? Be uncomfortable on purpose? It almost sounds un-American. All they've ever heard is: Plop, plop. Fizz, fizz. Fast Relief.

And in this case there is no fast relief. If you try, all you'll get is frustrated or fat. Frustrated—because the desire to smoke keeps coming back. Fat—because you'll try to eat the frustration away.

But why make yourself miserable trying to get rid of the desire to smoke? It's going to get rid of itself.

The discomfort is temporary.

The hardest phase, once withdrawal begins, lasts only two or three days. If you don't know any better, you'll believe that it's going to last forever. You'll think, "If this is what not smoking is going to be like, I can't live my life this way." And you'll bail out before you have a chance to see what it's like on the other side.

The first couple of days *are* uncomfortable. But addicts have a tendency to exaggerate. You convince yourself that withdrawal is something horrible. And if you're "dying" for a cigarette, it's the perfect justification for shooting up your drug.

On Thursday night, 24 hours after they stop smoking, I ask my class to list the physical symptoms that they're going through. If they've learned how to avoid making themselves feel deprived, angry and uptight, they don't talk about being in pain or climbing the walls. The most they come up with are symptoms like blurry vision or dry mouth, perhaps dizzyness, shakiness, perspiring or feeling tired.

George complains about having a stomach ache. Mike and Lisa both had trouble sleeping. Jackie says she feels like she has a mild case of the flu. Her joints ache and she feels slightly queasy. She can't think straight and drops things in the kitchen.

Yes, it's inconvenient and uncomfortable to stop smoking. But where's the agony? Since they haven't put themselves in the artificial prison of deprivation, they don't feel like they're trying to claw their way out. It may be annoying not sleeping normally for a night or two, or not being able to concentrate at work for a couple of days. But minor discomfort isn't what drives people back to smoking and risking their lives.

GETTING THROUGH WITHDRAWAL

During the two or three days of withdrawal, you may experience some unpleasant physical symptoms, as well as mental and emotional turmoil. As you readjust emotionally, your moods can change unpredictably. Your tolerance for frustration will be lower than usual, and you will have repeated desires to smoke.

And sometime during this period, your withdrawal will reach a peak, which can last for a few hours or even half a day. Now your desire to smoke will become an obsession; this is the essence of withdrawal from nicotine. And if you don't go through this struggle, something is wrong.

When I talk to Bill on Saturday, three days after he's stopped, he says he's just fine.

"I haven't even thought about it. I don't have a desire to smoke at all."

"Where'd it go, Bill? Don't you think this is a little strange for somebody who's been smoking a pack and a half a day for almost 20 years?"

"Oh, don't worry about me. This is exactly what happened the other time, too. I stopped for six months, when my uncle died of throat cancer, and I didn't have any problem."

"And remember what happened last time? You went to your class reunion and the desire suddenly came back, didn't it?"

"Yes, it did."

"Let me explain something, Bill. You have the ability to repress your desire, to turn it off. You could be doing it unconsciously, either because of fear that you'll give in if you do have the desire, or because you're so excited about not smoking that the desire to smoke gets pushed down for a while."

"Are you telling me I should be worried?"

"No, but there's something I've seen happen over and over: 'Easy stop, easy start.' The desire to smoke will come back sooner or later. Something will trigger it, and at that point you'll have a real choice to make. You'd better be ready for it."

In typical withdrawal, desires to smoke come in waves, lasting anywhere from a few minutes to an hour or two.

As the wave builds, your desire to smoke gets stronger, and you feel an almost irresistible pull. Often, just as it hits its peak and you think "I'm never going to make it," the wave breaks, and the obsessive desire subsides. Now you feel good for a while, pleased that you're winning. As the next wave builds, you'll feel more confident, knowing it will eventually subside.

Don't expect to have your best day at work during this time. But you can navigate your way through it, just as you do if you're suffering from a bad cold or if you have a major problem on your mind.

Keep as close to your normal routine as seems reasonable, but focus on this task as your *highest priority*. Don't forget: you're struggling against a drug addiction: you're making a life-and-death choice.

CONFRONTING YOUR ADDICTION

Getting through withdrawal means confronting your addiction head on. Here are five steps that will help you work through the compulsion to smoke, not only during withdrawal, but at any time in the future. First, begin your thinking by identifying what is happening to you.

1. I'M WANTING TO SMOKE RIGHT NOW BECAUSE I'M ADDICTED TO NICOTINE.

Every time you have a desire to smoke, face it. The desire is going to come over you whether you like it or not. It's normal. That's what makes you a smoker.

But you don't have to be afraid of the desire. It's not bigger than you. It can't hurt you. You don't have to try to get rid of it, hide from it, or pretend it isn't there. If you take the time to stand up to all your justifications and addictive thinking, the desire will fade away.

You may worry that you'll never feel normal again, but you will. You can be certain that as long as you don't give in to the addiction, your desire to smoke will inevitably diminish, becoming less frequent and less intense until most of the time you feel like a non-smoker. And each time you want a cigarette, remember what you *don't* want: to become sick and disabled or to have to go on smoking for the rest of your life.

2. And, I CAN SMOKE. I'M NOT DEPRIVED.

Nobody's taking your cigarettes away from you. You don't have to give up smoking, and even if you do stop, you can go back to smoking anytime you choose. What you can't be, however, is a happy, comfortable smoker or a part-time smoker. It's all or nothing.

3. And, ONE PUFF AND IT'S ALL OVER. I WILL GO BACK TO SMOKING COMPULSIVELY, EITHER GRADUALLY OR INSTANTLY.

Don't trick yourself into believing you can have just one puff when the going gets difficult. Using your drug to help you withdraw from your drug doesn't make sense. One puff will always call for another puff, and sooner or later you'll be back to smoking them all.

4. SO RIGHT NOW, I HAVE A CRITICAL CHOICE TO MAKE FOR MYSELF. I CAN EITHER:

a) REFUSE TO STAND UP TO MY ADDICTION, GIVE IN TO IT, AND SMOKE AND SMOKE AND SMOKE COMPULSIVELY, AND WISH I COULD STOP AGAIN, OR I CAN:

b) WILLINGLY ACCEPT THIS TEMPORARY DIS- COMFORT AND HAVE:

Starting with the benefits relating to your health, list ten you consider most important. To make sure these are your best reasons for stopping, ask yourself: am I willing to accept the temporary discomfort of withdrawal to have these benefits? Are they worth fighting for?

You only have two options, so now make your choice.

5. AT THIS MOMENT I CHOOSE TO WILLINGLY ACCEPT THIS TEMPORARY DISCOMFORT, BECAUSE I WANT:

Name at least three benefits.

When you have a craving to smoke, don't let it make a fool of you. Use these five steps, along with your list of benefits, every time the desire to smoke comes up.

First, recognize what's happening; you're having a desire to smoke. And you can smoke; you're not deprived. Then, remind yourself that one puff will take you straight back to the slavery of smoking. Finally, make your choice.

Don't whine and complain because you can't have it your way—smoking without consequences. You have to pay a price; you can't have your cigarettes and your health, too. Remember that you are choosing between temporary discomfort and ongoing misery. You can succumb to your desire and give in to your addiction for relief from temporary stress. Or you can resist the urge for the sake of long-term happiness and health.

By using these steps to face your addiction, you will train your mind to respond to the desire to smoke in a new way. Instead of automatically giving in to your desires, like a robot, you will confront the desire and make a choice.

It's like using a choker collar and leash to train an unruly Great Dane. At first it will keep trying to get away from you to go its own way. But if you persist, it will soon become perfectly disciplined; you won't have to struggle with it at all.

Since breaking an addiction takes time and is uncomfortable, it's easy to become discouraged. So focus on what you're trying to accomplish, on what you want the most, such as health, peace of mind, self-respect, and freedom from fear and slavery. You're doing something wonderful for yourself; if you succeed, you'll be grateful for the rest of your life.

TURNING YOUR DESIRES AROUND

When you learn to handle your addiction in this way, you can turn the strong desire to smoke into a strong desire to stay off smoking. You don't have to feel helpless anymore. You can feel powerful, because now you have a powerful tool for fighting back.

After my class has learned this process, I review it with them the following evening.

"George, have you had any desires to smoke?"

"I sure have. They kept me up half the night."

"What did you do about them."

"I thought about it like you taught me. I must have gone through those darn steps more times than I could count."

"Good, George. So how's it working?"

"Well, okay, I guess. It's the longest I've ever gone without a cigarette in 50 years. I used to get all worked up after just a couple of hours, but so far I feel pretty calm about it."

"That's because you know you can smoke, but you'd rather breathe, right?"

"You bet."

When I ask Jackie how she's doing, she talks about how different this is from all her other attempts to stop. "I've been concentrating on what you've said about substitution, because before when I wanted to smoke, I always ended up eating. Now I'm using the five steps instead of junk food—and it works a lot better."

"When you have a desire to smoke, what are you telling yourself?"

"First, I face what's happening. I tell myself I'm having an urge to smoke because I'm addicted to nicotine."

"Okay. What then?"

"Then I remind myself that I can smoke. I'm not deprived. I've still got my cigarettes right here in my purse."

"And then?"

"Well, since I know I can't smoke just a little bit without getting hooked again, I tell myself I have a choice to make."

"And what choice do you have?"

"I can either give in to the desire to smoke and go back to 30 cigarettes a day and hate myself, or I can willingly accept being temporarily uncomfortable and have more energy, no bronchitis, no more chest pains, and no more coughing at night."

"Good. Then what do you do?"

"I decide to accept the temporary discomfort until it's over with. And so far, I haven't been trying to relieve it with food."

"Sounds great, Jackie." I'm about to move on when I see her hesitating. "Something else?"

"Well, I have to admit I'm sure tempted. I've always had a problem with food and I'm pretty worried about it."

"You'll be able to get through it, Jackie, if you just keep working as you have been. Focus on your benefits—*they're* your cookies. They're your rewards for the discomfort you're choosing to accept. Nourish yourself on each and every benefit you're getting from not having to smoke. We'll be talking more about substitution, later."

I turn to Norman next. "How's it going, Norman?"

"Just fine. Except my stomach is tied up in knots."

"What are you doing when you have a desire to smoke?"

"I'm just telling myself I'm not going to smoke. I just push it out of my mind."

"How do you do that?"

"I'm not going to take that first puff, understand?"

"You know how many times I've heard, 'I'm just not going to smoke?' And the next week they're smoking again. I'm trying to get you to do something different this time, not just use your same old willpower method."

"You don't trust me, do you?"

"Norman, I trust you to do what every junkie does who comes through this class—resort to your old thinking. And my job is to get you to take time out to treat your addiction. Why aren't you thinking the steps through?"

"I'm too busy. I don't have time to mess with all that stuff."

"Well, if you don't have time to mess with it now, it's going to mess with you later on. If you can simply push it away, you haven't been hit with the strongest urges yet. You haven't peaked. One of these days you're to get into a crisis, and you'll break down and smoke."

"No, I won't, because smoking just isn't an option for me anymore. I've had half my lung removed."

"That's exactly the way Bill felt after he saw his uncle die of throat cancer. But he started smoking again, anyway. You can still smoke, Norman, even if you know it's going to kill you. You can say you'll never smoke again, but I've got news for you. Just like Bill, you can change your mind. You came in here to learn something new—not use your same old methods. Have they worked for you before?"

"No, but I didn't have the same reasons before. I just came here to get a little extra reinforcement."

"Well, what you don't understand yet is that a little reinforcement isn't enough to break down a compulsion that's been running your life—for how long now, Norman?"

"Since I was fourteen."

"Okay. That's 26 years. You're deeply attached to your cigarettes. You've been using them to cope most your life. They've got a powerful hold on you and to break it, you've got to be well-armed. Don't underestimate the enemy. You've been in Vietnam, right? What did you do when you were being attacked? Just push it out of your mind? You had to fight back with everything you could get your hands on. So when you have a desire to smoke, take three minutes to think it through. Use all five steps, not just one. Try it, Norman, it works."

By the time I ask Stephanie how she's doing, I'm warmed up and ready to face anything.

"Are you still feeling a little resentful at work?" I ask.

"No, not really, and I'm surprised. I tried to stop when they first banned smoking from the newsroom a year ago. But I was so worked up over the new rules, I wasn't able to think about what I really wanted for myself."

"You were so busy feeling deprived that you couldn't think straight, right? All you could do was react. What are you doing now?"

"Well, for one thing, instead of getting mad at my boss for wanting me to stop smoking, I'm really mad at the cigarette companies. They're pushing their drugs on us and don't give a damn about the consequences."

"That's for sure."

"And this might sound strange, but even though I'm having a

lot of desires to smoke, I almost look forward to them. I'm not scared anymore, and I don't feel helpless. I know how to fight back. So now it's kind of a challenge each time the desire comes up, wondering what kind of trick it's going to play on me this time."

"Excellent, Stephanie. You're on the right track."

When I ask Mike how he's dealing with his desires to smoke, he says, "Well, I can't say I'm accepting it willingly."

"Then how are you accepting it."

"Begrudgingly," he says, his face reddening with irritation.

"Why do you want to stop smoking, anyway?"

"For my breathing."

"What's wrong with your breathing."

"Nothing really. But I'm sure it must be bad."

"Wait a minute. What happens when you walk up a flight of stairs. Do you feel short of breath."

"Nope."

"You never feel short of breath?"

"Not that I notice."

"Then, Mike, that's *not* one of your reasons for stopping smoking."

I turn to the class. "How about you, George? When you walk up a flight of stairs, do you feel short of breath?"

"You'd better believe it. I feel short of breath when I walk across the room."

"Okay, that's George's reason for stopping smoking. So, Mike, what's yours?

"Well, it's other people. It's my wife and my daughter. And my son. I've been after him to stop smoking pot, and he tells me, 'As soon as you stop smoking cigarettes, Dad'."

"I see. No wonder you're not willingly accepting the discomfort. I tried to make it clear at the beginning of this class that you can't stop smoking to please other people. So what are you doing here?"

There's a long pause before he replies. I see Vera squirming in her chair, upset at someone being put on the spot.

Finally, he says, "I guess it would please me, too."

"How's that? How does it make you feel to smoke around your daughter?"

"I feel lousy. I feel like a bad example. When my daughter comes home from college, I go into the garage to smoke."

"Do you mean that you're tired of feeling like a hypocrite because you tell your kids one thing and you do another?"

"You could say that."

"What about your wife? How does that affect you?"

"Well, she's allergic to smoke, and I feel guilty about it. I think smoking comes between us."

"Okay. You're not stopping smoking because of your breathing. That's not your problem—yet. It's the guilt about your wife's health and the remorse about how you're influencing your children. What you want is to have pride and self-esteem and to not have this conflict in your life. Does this sound like something you'd willingly fight for?"

Mike seems to loosen up for the first time in four days. "Well, lady, I guess you hit a bingo on that one. I can see why I've never made it before, because I thought I was doing it for them and not for me. But it *is* for me.

Vera sighs with relief. She's starting to see that you don't have to be afraid of having your thinking challenged. In fact, this is often essential.

SEEKING SUPPORT

Without coming to a class for help, Mike would have a tough time sorting out the mistakes in his own thinking—and he wouldn't be able to work through deprivation and stay off smoking. And Bill, who's so good at repressing his desire, wouldn't be prepared to deal with it when it finally surfaces.

Because George and Vera have smoked for most of their lives, they're going to need a lot of support in the beginning in order to see that there is "life after smoking."

All four of them need help "tracking" their own thinking—understanding how thoughts trigger feelings and lead to certain actions. They need someone to listen to them and play back their words, pointing out the errors, so they can clearly see how their thinking goes astray. They're learning a new skill, and it's so much faster and easier with the help of an experienced coach.

While many people will be able to stop smoking successfully by reading this book, some may need additional help for the next phase of their recovery—coping with the normal ups and down of life without nicotine.

This book has everything you need to understand the nature of your addition, but even the best book may not be sufficient to help you face the pitfalls and unexpected complications in this your phase of recovery. What happens if your relationship breaks up, if you get fired from your job, if you fall into a severe depression, if a loved one dies? Most of us need support to get through such situations—counseling or a self-help support group.

You also need to continue learning the skills outlined here in order to avoid defeating yourself. That means taking the time to think through the five steps every time you have a desire to smoke. Your addictive thinking is a bag of tricks, ready to con you out of all your hard-earned freedom. Upsets and crises can make you want to smoke, but they can't make you go back to smoking. In order to take that first puff, you have to find some way to talk yourself into doing it.

And like all drug addicts, you're very good at that.

"You know what? I just realized something," Ellen
says with disbelief. "I planned my whole relapse!"

CHAPTER

6

JUNKIE THINKING

PART I: DEFENDING YOURSELF AGAINST SELF-SABOTAGE

If you hope to recover from nicotine dependency, you need to be aware that your addictive thinking will work against you unless you learn to defend yourself against it.

Most stop-smoking methods stress the importance of avoiding temptations around you: coffee, alcohol or spicy foods; parties, stressful situations, even sitting in your favorite chair. It's true that when you stop smoking, many things in your daily life will trigger the desire to smoke. When this happens, take time out and work through the desires, using the steps from Chapter 5. But the real danger is within: the thinking created by your addiction to cigarettes—junkie thinking.

Like believing you can't smoke, junkie thinking is a form of self-deception. You use irrational, distorted or irrelevant arguments to justify taking that first cigarette. And along with feelings of deprivation, it's the most common cause of relapse.

Most of us are disarmed in the face of our own irrational thinking. We become as helpless as little children. It doesn't even occur to us to question the false assumptions which sabotage our best efforts.

"In the past, when you've tried to stop smoking, how did you talk yourself back into it?" I ask my class. "Lisa, can you give me an example?"

"Yeah, I remember the last time I stopped, I started to think that maybe smoking wasn't so bad, after all."

"That's one of the difficulties of trying to stop at a young age, Lisa. It's easy to think that you've still got a lot of time to deal with it. But you've got good reasons to stop *right now*, don't you?"

"Yes. I have asthma."

"That's right. So here's the first thing you need to do. Point out the junkie thinking: 'smoking wasn't so bad.' Then, ask yourself, what's wrong with that thought? Challenge it. Turn the words around, so that you make them work *for* you, rather than *against* you. Then you'll be talking yourself out of taking that first cigarette, instead of talking yourself into it.

"Talk back to your addiction. 'Of course the smoking wasn't so bad. You liked the smoking. But did you like the consequences? Remember that your breathing was bad. Worrying was bad. Being afraid you might suffocate in your sleep was bad.' Would that work for you, Lisa?"

"Yes, that makes sense," she says. "It's true. I like to smoke, but I don't like the consequences."

"Good. What about you, Ellen? How are you doing this time?"

"Oh, much better. I didn't say anything about it the last time I stopped, but I was really envious of other people when I saw them smoking. And, I guess, I was harboring doubts about whether I was really addicted."

"So, what were you thinking exactly?"

"I was thinking, 'They get to smoke. They're enjoying it. Maybe one wouldn't hurt me'."

"And what did you learn, Ellen?"

"That if I smoke one, I have to smoke them all."

"Absolutely. Whatever you do, keep reminding yourself of that. How about you, Bill? Any junkie thinking in the last couple of days?"

"Well, I'm not sure what kind of thinking this is, but I've been wondering whether I really want to stop."

"Sounds like junkie thinking, all right. Junkie thinking is *whatever justification or rationalization you use to make it okay to take that first puff.* Bill, as a good lawyer, what's the first thing you need to do when you get a new case?"

"Find out as many facts about the case as I can."

"Right. Get the facts straight. So when you start wondering whether you really want to stop smoking, the first thing you need to realize is this: you have stopped, haven't you?"

"Since yesterday, anyway."

"Then, the correct question is, 'Do I really want to stay off smoking?' You need to ask the right question to get the right answer. If the answer turns out to be 'No, I don't really want to stay off,' then what you're saying is, 'I really want to go back to smoking 30 cigarettes a day.' Is that what you want?"

"That's not what I want."

"I didn't think so. Bill, you know how to stand up for other people. So, to protect yourself, you need to become your own defense attorney. We all do. We need to stand up and defend ourselves against the lies and distortions of junkie thinking.

"Mike, can you think of anything that might trip you up?"

"Well, the last time I stopped, I don't think I was really doing it for myself. And one Sunday afternoon, when I was alone, I decided to smoke just one, because nobody would know about it."

"Oh, I get the picture, Mike. There you are, everybody's gone, and suddenly you think, 'They've left me here alone! I could just go out into the garage, have one little cigarette and nobody would ever be the wiser. Who'd know?' So you go out into the garage, light up and smoke your one cigarette, and you're just delighted with yourself because you're getting away with something. Isn't that fine. So what's wrong with this picture? Who'll know if you smoke?"

"I would," Mike says.

"So what? I'm sure you have your secret little peccadillos that only you know about. That's not the problem. Come on, who'd know?" I ask again.

There's a long pause. Finally, Stephanie says, "Everybody."

"That's right. Why? What happens when Tuesday night rolls around, Mike. Your wife is at the PTA meeting, your kids are at the movies, and you realize you're all alone again. And there you go—back to the garage. Before you know it, you're smoking two packs a day. Is that the way it happened?"

"Pretty damn close. You must have been there."

"No, this one's a classic. And you don't have to fall for it anymore, Mike. If you sneak one little cigarette that nobody knows about, you can be sure that you're going to try to get away with it again and again. Tell yourself, 'Sooner or later, everybody will know, because I'll be back to smoking. Sneaking one little cigarette will never be enough for me'.

"It's like a petty criminal who tries to get away with one burglary and ends up in jail," I tell the class. "He hates it in there and swears this is the last time. But once he's out, he spots the perfect opportunity to pull one off, and the next thing he knows he's back in jail again."

I notice that Vera is getting a little edgy. "What's going on with you, Vera?"

"I just remembered something I thought this morning."

"What was it?"

"Oh, I don't want to talk about it. It's too embarrassing."

"Go ahead. All of us have these crazy little thoughts. It's normal."

"You aren't going to believe this, but I thought, if somebody died, then I could smoke. Isn't that awful?"

Even though the others laugh, they're a bit taken aback.

"Isn't it amazing, Vera, the lengths we'll go to, to make it okay to smoke? Here's how I'd answer that thought: 'Who does it have to be? Who has to give up their life, so that I can feel justified in smoking? Nobody. I don't need a crisis to blame my smoking on'."

"It sounds like Vera's feeling deprived to me," Bill says.

"You're right. And isn't it interesting how subtle deprivation can be? You feel like you can't smoke until there's a big enough crisis to justify it. But you can smoke anytime you choose. You *do* have to live with the consequences, however."

THE PERFECT PLAN

Those of my clients who've gone back to smoking often describe it as something that "just happened." But it never works that way. They never intend to go back to smoking. But they've found some excuse to take that first puff, some way to talk themselves into using their drug. Yet they don't want to admit it. Junkies lie to themselves; they don't want to take responsibility for their own decisions and actions.

While reviewing the pitfalls of junkie thinking, Ellen asks about being sabotaged by the subconscious, about going back to smoking without really thinking about it.

"Is that what happened to you last time, Ellen?"

"Yes. I was very busy at work and trying to sell my house at the same time, and in a moment of tension I grabbed a pack of cigarettes I found lying around, and I just smoked one."

"Why didn't you think the desire through?"

"I don't know. I just went on automatic pilot and did it."

"All right. You're saying you didn't think at all. But wait a minute. A little while ago you were telling us that every time you saw someone smoke, you were thinking how nice it looked, and that maybe you could have one without going back to smoking."

"Oh, yes, that's true."

"You don't remember what you thought at the time you smoked, but you'd been doing a lot of junkie thinking *before* that.

And you weren't challenging it, were you?"

There's a pause before she answers. "You know, I just realized something. I can see now that I planned my whole relapse."

"You did?" I say, taken by surprise. "How?"

"Well, I remember now that two weeks before my house was sold, I was telling myself, 'If my house doesn't sell, I'm going to have a cigarette.' Of course, my house did sell—but then I got a huge bill from my attorney. So I decided to smoke over that."

I turn to the class. "How about that? She planned the whole thing and then promptly forgot she did it. But, Ellen, tell me, after all you had learned here in class about addiction, weren't you concerned that if you smoked the first one you'd go back to smoking?"

"I was worried about it. But even during class I thought you were exaggerating about smoking being a drug addiction. I guess I was still in denial about it. I even thought the intense follow-up work you do by phone wasn't really necessary in my case. When I'm counseling other people, denial is always so obvious, but it's harder to spot in myself."

PART II: TREATING ADDICTIVE THINKING

The more you take the time to work with your junkie thinking, the better you get at dealing with it. If you train yourself to talk back to these self-sabotaging thoughts, you will neutralize them. Eventually they will lose their power over you.

The process for treating addictive thinking works like this. First, identify the junkie thought. Next, respond to it for what it is—a lie. Give yourself some evidence why this thought is false, and tell yourself the truth. Finally, since junkie thinking is a sure sign that you're wanting to smoke, review the five steps from Chapter 5 and make your choice.

Below are some of the common arguments you might use to

rationalize your way back to your drug. If you're not prepared to fight back, cigarettes can make a fool of you, even months or years after you've stopped smoking.

The process of relapse almost always begins with the thought that you can get away with smoking *just a little bit*. You may hope you can control it this time, because you want to be able to smoke like a social smoker. Too bad. You don't get to. You're addicted.

If you believe you can stop at one cigarette or control your smoking by becoming a social smoker, you are still refusing to accept that you have a drug problem. And like Ellen, you will only learn by the painful experience of relapse.

DENYING YOUR ADDICTION

The following four examples are not only dangerous in themselves, but prop up the rationalization behind most of the others.

❖ **Junkie Thinking**: "One puff won't hurt."

Response: "One puff has always hurt me, and it always will because I'm not a social smoker. One puff and I'll be smoking compulsively again."

❖ **Junkie Thinking**: "I only want one."

Response: "I have never wanted 'only one.' In fact, I want 20 or 30 a day every day. I want them all."

❖ **Junkie Thinking**: "I'll just be a social smoker."

Response: "I'm a chronic, compulsive smoker, and once I smoke one I'll quickly be thinking about that next one. Social smokers can take it or leave it. That's not me."

❖ **Junkie Thinking**: "I'm doing so well, one won't hurt me now."

Response: "The only reason I'm doing so well is because I haven't taken the first one. Yet once I do, I won't be doing so well anymore. I'll be smoking again."

THE JUNKIE PLAN

The first time you go a couple of weeks without smoking, it's easy to con yourself into believing you're cured. There is no cure for addiction.

If you do admit that you're addicted and that one puff will always take you back to smoking, you may devise a junkie plan such as this:

❖ **Junkie Thinking**: "I'll just stop again."

Response: "Sounds easy, but who am I trying to kid? Look how long it took me to stop this time. And once I start, how long will it take before I get sick enough to face withdrawal again? In fact, when I'm back in the grip of compulsion, what guarantee do I have that I'll ever be able to stop again?"

Saying you'll do it later is a method you use to avoid the discomfort now. And if you won't face it now, why do you think that you'd be willing to face it later?

❖ **Junkie Thinking**: "If I slip, I'll keep trying."

Response: "If I think I can get away with one little 'slip' now, I'll think I can get away with another little 'slip' later on."

Think about what it really means to slip. It's an accident, something you would avoid if you could, like falling on an icy sidewalk. You're walking along and suddenly your feet shoot out from under you. There's nothing you can do about it. You have no choice.

When you give in and smoke, however, you make yourself feel better by pretending you didn't really have a choice. Also, using the word "slip" is a way of trying to minimize the importance of the choice you're making, and a way of avoiding the responsibility for that choice. In this case, a little slip will have dramatic results. It's like "slipping" out the window of a ten-story building. The truth is, when you're wanting to smoke and you take that first puff, you're making a deliberate choice to do so.

❖ **Junkie Thinking**: "I need one to get me through this withdrawal."

Response: "Smoking will not get me through the discomfort of not smoking. It will only get me back *to* smoking. One puff stops the process of withdrawal and I'll have to go through it all over again."

When you're at the peak of withdrawal, you're close to a breakthrough. It's absurd to calm yourself down with a cigarette when what you really need to do is accept the temporary discomfort and work your way through it.

SELF-PITY

❖ **Junkie Thinking**: "I miss smoking right now."

Response: "Of course I miss something I've been doing every day for most of my life. But do I miss the chest pain right now? Do I miss the worry, the embarrassment? I'd rather be an ex-smoker with an occasional desire to smoke, than a smoker with a constant desire to stop doing it."

❖ **Junkie Thinking**: "I really need to smoke now, I'm so upset."

Response: "Smoking is not going to fix anything. I'll still be upset; I'll just be an upset smoker. I never *have to have* a cigarette. Smoking is not a *need*, it's a *want*. Once the crisis is over, I'll be relieved and grateful I'm still not smoking."

When talking to smokers, I often hear this refrain: "I stopped once and I was doing okay until . . . my car broke down . . . I had a fight with my husband . . . my cat got lost. . . and then I just had to have a cigarette." You may even manufacture a little crisis so that you can have a good excuse to smoke. Remember, a lot of things can make you *want* to smoke, but nothing can make you smoke.

❖ **Junkie Thinking**: "I don't care."

Response: "What is it exactly that I think I don't care about? Can I truthfully say I don't care about chest pain? I don't care about gagging in the morning? I don't care about lung cancer? No, I care about these things very much. That's why I stopped smoking in the first place."

Sometimes you may *feel like* you don't care, but it isn't true. Sometimes you *wish* you didn't care, because it's easier to drug yourself than to cope with your feelings. Too bad, because you really do care.

❖ **Junkie Thinking**: "What difference does it make, anyway?"

Response: "It makes a difference in the way I breathe, the way my heart beats, the way I feel about myself. It makes a tremendous difference in every aspect of my physical and emotional health."

When you're depressed, you may feel that nothing matters. But if you really believe nothing matters, why don't you throw your television set out the window, wreck your car and burn your house down? You don't, because those things have value, and you know that the depression is temporary. Then why throw freedom from smoking out the window, as if your health were less valuable than your television?

❖ **Junkie Thinking**: "Why bother? We've all got to die sometime."

 Response: "I'm alive now, and I want to live as well as I can for as long as I can."

Of course, we've got to die sometime, but how do you want to live in the meanwhile? Dragged out, sick, gasping for breath? Hooked up to an oxygen tank for years? And if you don't care about dying, why bother to fasten your seat belt or look before crossing the street?

❖ **Junkie Thinking:** "I deserve to smoke."

 Response: "I deserve to be free from the pain and fear and self-disgust of having to smoke. I do not deserve the miserable consequences of using drugs as rewards."

If you've used cigarettes to reward yourself for getting through life, you'll be a sucker for this twisted argument. Once you've stopped for a few days, or even hours, you may convince yourself that you deserve to smoke as a reward for not smoking. Benefits are *your* reward for not smoking.

❖ **Junking Thinking**: "Cigarettes are all I've got left."

 Response: "Nonsense. I know I have other things to live for. And could my drug addiction be standing in the way of what

I really want, such as my health, peace of mind, and the energy to make positive changes in my life?"

If you believe that your drug addiction is the only thing that makes your life worthwhile, maybe you need to ask yourself, how did this happen to me? How have I allowed my life to be reduced to this sad and pathetic state?

You may have experienced tragedy and loss in your life, but is that an excuse to spend the rest of your life holed up with your drug, refusing to work through your pain and move on?

DEPRIVATION

❖ **Junkie Thinking**: "They get to smoke."

Response: "I can smoke just like they are, but then I have to smoke like they do, constantly and compulsively."

From time to time you may envy people lighting up at the next table. If you think, "They get to smoke," you're implying, "They can, but I can't." This is a lie. You can smoke, and allowing yourself to feel deprived will only lead to self-pity and defiance.

GLAMORIZING

❖ **Junkie Thinking**: "It would taste so good."

Response: "Come on, most of the time I didn't even notice the taste. And one thing is certain, I wasn't killing myself for taste."

Don't glamorize smoking—how wonderful that cigarette looks, how great it smells, or how good it would taste. You don't need to reinforce your addiction. The cigarette companies do plenty of that already. At any rate, what you really mean is that a cigarette would sure "feel" good right now, because it would relieve your craving to smoke.

❖ **Junkie Thinking**: "They're smoking and it's not hurting them."

Response: "Could anyone have known how I felt simply by watching me smoke? Did I make awful faces when it burned my throat? Did I announce my chest pain or my fear when I had heart palpitations? Other people only saw the cool, glamorous image I tried to project. It was an illusion, just like the one other smokers are projecting."

Just because people don't look sick, don't think smoking isn't hurting them. The pain and the fear don't show.

❖ **Junkie Thinking**: "Smoking wasn't really so bad."

Response: "If it wasn't so bad, why have I been wanting to stop and trying to stop for so long? How bad does it have to get?"

If you're letting yourself forget how bad it was, *read your remember letter.*

FALSE OPTIONS

❖ **Junkie Thinking**: "If I don't eat something, I might smoke."

Response: "If I don't give in and eat right now, I may feel uncomfortable, but it won't make me smoke. The discomfort I'm feeling is temporary and will never be too hard for me to get through. I don't have to stuff my feelings with food."

Stopping smoking is uncomfortable, and you may think that if you don't distract yourself with food, you'll give in to smoking. The choice is not between eating and smoking. Your choice is either to give in to your addiction or to work through the withdrawal.

❖ **Junkie Thinking**: "If I don't smoke, I might drink."

   Response: "Not smoking will not force me to drink. If I'm having desires to drink, I need to deal with that problem, not avoid it by giving in to smoking."

Trying to kick one drug can bring on the desire to use another one. If you're a recovering alcoholic, you naturally make staying sober your top priority. In a crisis or emotional upset, you may try to justify smoking by saying, "If I don't smoke, I might drink. I'll choose the lesser of two evils." A recovering heroin addict could use the same crazy logic: "If I don't drink, I might shoot up."

Remember that the choice is either to accept the temporary discomfort of change or to drug yourself—to learn how to cope with life's problems or to run away from them.

HOW CRAZY CAN IT GET

❖ **Junkie Thinking**: "This won't count because . . ." (It's not my brand; I'm on vacation; I'm drunk; I'm angry and I know what I'm doing).

   Response: "The first puff counts no matter when, where, why or how I do it. The circumstances don't alter the reality of my addiction."

❖ **Junkie Thinking**: "I know I can't smoke just one, but I hope I can."

Response: "If I'm still *hoping* I can get away with it, then I'm kidding myself. Telling myself I know something is true and hoping it isn't, is totally irrational. If I don't get smart, I'll have to learn the hard way."

❖ **Junkie Thinking**: "Do it fast before you think about it."

Response: "Too late, I'm thinking about it right now. If I go ahead and give in, I'll be making a deliberate choice to do so. I'd better think it through and make my choice."

THINK BEFORE YOU ACT

Think before you act. You're not a cigarette processing machine for R. J. Reynolds & Co. You're not a robot, even though you've been acting like one for years.

Because you've smoked for so long, you're programmed to produce lots of junkie thinking. It can come fast and furious. For example, once you fend off one junkie thought, you may be hit with another one or two. In a crisis, you tell yourself, "I need a cigarette." If you successfully counter this with, "Smoking won't fix anything," your mind jumps to, "*Yes, but* one won't hurt." If you're able to turn that one around by remembering that one always leads back to smoking, yet another trap opens, "*Yes, but* you can always stop again."

If you're not prepared to bring yourself back to your senses, later you'll be wondering, "How could I have been so foolish?" Follow through and respond each time your mind comes back with, "Yes, but . . . "

The compulsion to use a drug creates a kind of temporary insanity. The craving is like a haze that comes over your mind. It clouds your reasoning ability, effectively lowering your intelligence for the moment. Why else would someone who is otherwise bright and responsible break down so easily?

Don't underestimate the power of junkie thinking. Don't treat it lightly. Get tough with it.

Defend yourself against absurd ideas like "I'll just have one puff." If possible, talk back out loud, "Oh, bull! I know one puff is never enough for me." Use sarcasm or your favorite profanity— whatever works best.

No matter what happens, you never *have* to take that first cigarette. With sufficient effort, you can always make the decision not to smoke. And your health and your life are worth any effort you make.

Jackie didn't even like pretzels, but the day after she stopped smoking, they suddenly seemed irresistible.

CHAPTER
7

STUFFING IT

SMOKING FOOD:
DETOUR TO DISASTER

Many smokers, and women especially, are terrified about gaining weight; they expect it to be an automatic side effect of giving up cigarettes. Often, they're already fighting those extra pounds while they're still smoking. They're so worried, they'd prefer to go to the grave smoking rather than face the prospect of an additional ten or 20 pounds.

And if they go for help to a stop-smoking program, they get this advice. "You'll need substitutes for the pleasure of smoking," a typical pamphlet says. "You can get oral satisfaction from low calorie snacks, gum, or toothpicks. You can keep your hands busy with finger puzzles, pencils, or coffee stirrers. Have a survival kit ready for the moments you'll need them."

Another brochure suggests: "Miss having something in your mouth? Try a toothpick or a fake cigarette." Or: "Find a substitute for cigarettes. Reward yourself with lemon drops, popcorn, carrot sticks, whatever."

As I'm reading these excerpts in class, Stephanie suddenly

interrupts me. "I went to a program like that! On the table in front of us were straws, sunflower seeds and hard candies—and the man who was teaching the class was 50 pounds overweight. Needless to say, I didn't go back."

Many stop-smoking programs teach exactly the opposite of what you need to know in order to keep from gaining weight when you stop smoking. Compulsive smoking isn't a problem of having something in your mouth. You just happen to use your mouth to get nicotine to your brain. Treating smoking as an oral fixation is as ridiculous as it would be to treat cocaine addiction as a nasal fixation.

"In other words," I point out in class, "if you're hooked on cocaine, it's something to do with your nose, right? So why not snort baking powder instead?"

There is no substitute for cigarettes. And when you try to find one, you end up eating an awful lot of food. What starts out as a nice idea, munching on healthy, low-cal carrot and celery sticks, soon degenerates into an obsession with corn curls and chocolate chip cookies.

Since replacement only reinforces hand-to-mouth behavior and avoids tackling addiction, there's a good chance you'll relapse and end up a fat smoker. The only reason people try to find substitutes for cigarettes is that they don't know what else to do.

DON'T RUN, DON'T HIDE

You don't have to gain weight when you stop smoking. Here's how you can avoid it.

❖ First, write down what you normally eat as a smoker. What do you usually have for breakfast, mid-morning snack, lunch, after-noon snack, after work, dinner, and evening snacks? This will give you something to use for comparison.

❖ Second, once you stop smoking, keep a food diary for the first few days. Write down everything you eat, how much of it you eat, and when you eat it.

❖   Third, look for problem times that you know will be difficult for you—for example, breaks at work, mealtimes, or watching TV in the evening.

When the problem times arrive, before you eat anything more or other than you normally would, ask yourself, "If I were still smoking now, would I be wanting this food or would I be reaching for a cigarette?" If the answer is, "I would not usually be eating this," treat it as a desire to smoke and choose to accept the temporary discomfort rather than give in to it. If you don't make this choice, you*r compulsive desire to smoke can transform itself into a compulsive desire to eat.*

Even if it's normal for you to overeat, don't try to go on a diet. Take care of your smoking problem first. Tracking your eating habits will keep you from gaining any more weight. You may even lose a few pounds.

Losing weight, however, is not the goal at this time. *Not gaining* is a big enough task for now.

❖   Fourth, ask yourself, "How am I feeling right now? Am I bored, angry, lonely, worried? Have I always smoked this feeling away so I can tolerate the situation better? Am I wanting to fill in an emptiness or numb this feeling with food?

❖   Fifth, if you discover a feeling, respond to it appropriately. Cry, pace around, call a friend, get some help, or wait for the feeling to subside. Feelings are temporary. They come and go. If an uncomfortable feeling does not subside—if it keeps calling for medication such as nicotine, sugar, or alcohol—it's a sign that you may have some other problem you need to work on.

BLAMING YOUR BODY

When you stop smoking, you may deny, even to yourself, that you're eating more than normal. You will think of yourself as a *victim* of weight gain, not a perpetrator, because if something is happening *to* you, you don't have to take any responsibility for it.

The most common excuse is to blame any additional weight on changes in metabolism. It's true that when you stop poisoning yourself, your body assimilates nutrients better. And without the stimulant, nicotine, your heart rate slows down and you burn fewer calories. But since you will also have more energy, you're likely to be more active. And if you are, this change in metabolism will not cause you to gain much weight.

Recent studies show that changes in metabolism should cause you to gain no more than one pound a month. Therefore, it would take ten months to gain ten pounds. Most ex-smokers who gain weight, however, put on five, ten or even more pounds in the first few months. So if you gain an extra 15 or 20 pounds, don't blame it on your metabolism—or the bathroom scale. Take a look at your eating habits. Something has changed.

BREAKING THROUGH DENIAL

Whether they're trying to stop smoking or not, many people have problems with their weight. So special follow-up classes on substitution are very helpful. A few weeks after they stop smoking, Jackie and Stephanie attend one of these classes.

I begin by asking if anyone is having trouble with weight gain.

"I think I've gained about seven pounds," Stephanie says, "and I don't understand it."

"Have you been doing any substituting?" I ask.

"Well, I don't think I have. I've been chewing more gum. But mostly, I've been eating a lot of salads."

"And you haven't been snacking more?"

"I don't think so."

"Well, you may have gained a few of those pounds from changes in your metabolism, but you can counteract that. Now that you have more energy, you can get started on your aerobics classes. But be careful about snacking; it can sneak up on you. Isn't that true?" I say, turning to Jackie.

I know she's had trouble with substitution in the past, so I ask Jackie if she can recall in any detail what happened the last time she stopped smoking.

"I sure do," she says. "After I'd gained that 16 pounds and went back to smoking, I realized what I'd done. I'd gone to one of those programs that provide you with all that stuff to chew on, and the first night it was okay—I just nibbled on some healthy snacks.

"But during the morning break at work the next day, I had a donut with my coffee, instead of my usual cigarette. After lunch, I suddenly decided to run out and get a candy bar. That afternoon was pretty stressful, so I chewed gum for a while and polished off a baggie of carrot sticks I'd brought to work. But finally, I went out to the machines in the lobby for some cheese crackers— they're not too bad for you."

"Right. Just pacify your addiction," I say. "Throw it a little fat, sugar and salt—and hope it'll leave you alone. But was that the end of it?"

"I wish. After work, when I got home, I remembered we had some pretzels left over from the weekend. I don't even like pretzels, but all of a sudden they seemed irresistible. Later, I remember my husband asking me why I was having an extra cocktail before dinner, and I reminded him that I'd stopped smoking, and I needed something to calm me down.

"I know I ate more than usual at dinner, and then, after the kids were in bed and we were watching TV, I was still feeling a little antsy, so I asked my husband if he'd like some ice cream. He said he didn't care for any, but I decided to get some for myself anyway. And all this happened on the first day."

"Wow," I say. "You must have started worrying about gaining weight."

"No, not really. I figured I'd just get myself through the first few days, and then I'd be over it."

"Wouldn't it be nice if it were that simple."

"Yes, but I became obsessed with food. I'd decide I wasn't going to eat any snacks today, and then I'd find myself having just one more bag of M & Ms. I'd think, 'Do it and be done with it.' But it was never done.

"Finally one day I was getting ready to go to a wedding, and I realized there wasn't a thing I could wear, so I ended up not going. That's when I got furious and went straight to the store for some cigarettes."

"After you went back to smoking, Jackie, did you lose the 16 pounds?"

"Only part of it, so now I have two problems. I sure don't want to gain any more this time around."

"And how's it going this time?"

"Fine so far. I'm keeping track of everything I eat. And I'm not letting myself get away with a thing."

"Good. People don't understand that if they feed their addiction, it'll keep demanding more. It's like a stray cat that meows at the back door. You put a bowl of milk out and then you wonder why this cat keeps coming around. If you stop feeding it, it'll go away."

Since I noticed Stephanie taking some notes while Jackie was talking, I ask her whether she's thought of something new.

"Well, I don't know how I could have forgotten this, but I just realized that until a couple of days ago, I was having ice cream every night. I never used to do that. And I never used to have french fries with my sandwich at lunch. I've been doing that every day, too."

"Stephanie, what happened to all those salads you were telling us about?"

"Oh, I *have* been doing that—but just the last few days. I

started eating salads because I couldn't figure out why I was gaining weight. Until now, I'd just blanked out the other stuff."

"Isn't denial amazing? You can do things and not know you're doing them, because you don't *want* to know."

POOR ME

Teaching ex-smokers to avoid overeating means fighting a lifetime of conditioning. From early childhood, you learned how to use food to console yourself when you were unhappy, to numb uncomfortable feelings, and to reward yourself for your accomplishments.

From an early age, you've probably been given treats as consolation for life's disappointments. If you fell down and scraped your knee, lost something you loved or didn't get the part in the school play you wanted so badly, you learned to expect goodies as a way to feel better. If mother and dad didn't have the time or inclination to help you through your problems, they always knew how to console you with a cookie.

During drug withdrawal, it's natural to feel uncomfortable, upset, and a sense of loss. If you don't know that these feelings are temporary and that they will subside naturally, you'll look for something to make them go away. Letting go of cigarettes will take a little time. Eating to console yourself will take a lot of food.

VICTIM'S COMPENSATION

Choosing between cigarettes and health is a choice smokers don't want to make. You want to have both. When you can't have it the way you want it, you may feel deprived and resentful and look for compensation. It's common to have this junkie thought: "If I can't smoke, at least I can eat."

But of course you *can* smoke, and avoiding feelings of

deprivation will make it easier to keep from overeating. Tell yourself: "I can smoke cigarettes. I don't have to smoke food. My choice is not between smoking and cookies. It's between having the benefits I want from not smoking and going back to 40 cigarettes a day. I can either accept this temporary discomfort now or go back to being miserable every day."

REWARDING YOURSELF

One way to get a seal to jump through a hoop is to throw it a fish. When people stop smoking, they tend to act like seals, expecting food for their efforts. Or they act like children, who clean their rooms or do the chores not to please themselves but to please their parents—and to get a reward. You tell yourself, 'I haven't smoked today. I've been good, and like a good child I deserve a reward: potato chips, fudge, German chocolate cake.'

When you stop smoking, you're making an adult choice. You're doing it for your own happiness and well-being. You need to nourish yourself on the real rewards. You're not coughing or worrying. You can breathe. You've got freedom and peace of mind. What do Twinkies have to do with it?

JUNK-FOOD JUNKIES

Lately, in some of the biggest mergers in corporate history, cigarette companies have been buying up food companies.

"What's behind this diversification into food?" I ask my class. "Have you wondered? These companies know your weaknesses. They know how easy your compulsion to smoke can switch over to a compulsion to eat. And they're going to be ready for you with all those yummy goodies."

THE DEADLY DIET

People sometimes use smoking to suppress normal hunger. Since nicotine cuts down appetite, you may feel more real hunger when you stop smoking. But you'll also have more energy and a desire to be more active, so you'll burn up more calories. Satisfying true body hunger doesn't normally make people fat; using food for other purposes does.

By killing normal appetite, smoking can be used as a way to control your weight. "I never eat breakfast or lunch. I just have coffee and cigarettes," is something I've heard many times, especially from women.

It's a deadly diet—controlling your eating through chronic poisoning.

Yet the appeal of this "diet" explains the huge success of cigarettes like Capri and Virginia Slims. Their advertising preys upon the fears and anxieties of women—through the powerful combination of slogans, brand name, images of sleek and slender women, and the long, thin shape of the cigarette itself.

STUFFING FEELINGS

If you're like most Americans, you've been taught to seek immediate relief from discomfort or pain. If you have a headache, take an aspirin; if you're tense, have a drink. If you have an upset stomach, don't question the kind of food you're eating, take an Alka Seltzer. Tired? Don't figure out how you can get more rest, just drink more coffee.

Rather than getting to the source of the problem and finding a solution, you're encouraged to look for an escape.

Many people who stop smoking are surprised by emotions they haven't felt in years. Without knowing it, they've been repressing their feelings with cigarettes. Now when sudden anger erupts, when resentment and depression sweep over them, they panic. They've never allowed themselves to feel this way. Not knowing how to cope, they treat these sensations the same way they treat a headache or an upset stomach. They reach for something to get rid of them.

Ellen told me about an experience she had right after she stopped smoking the first time she took my seminar. A soft-spoken and courteous woman, who prided herself on staying cool and in control, Ellen found herself suddenly enraged when she was caught in rush-hour traffic on her way home from work. She couldn't wait to get off the freeway and stop at a fast-food stand. Anxious and irritable, she devoured a hotdog—something she hadn't eaten in years—but was still on edge. She only began to calm down after a large order of fries and a strawberry milkshake.

Later she realized that she'd used cigarettes to suppress anger, ever since she'd been abused as a child and started smoking at the age of 15.

Once Ellen realized that her tendency to suppress feelings with food could become compulsive, she was careful to work through her feelings in moments of tension.

UNFORTUNATE TIMING

By the time many smokers realize that they need to stop smoking, they are facing mid-life changes and difficulties. Their metabolism is naturally slowing, and their lifestyle has become more sedentary, causing them to gain weight. At the same time, they're having problems with careers, relationships, children, and personal identity.

Because of this unfortunate combination of circumstances, stopping smoking not only requires an understanding of this complexity, but often requires extra support through counseling, relaxation training, or self-help groups.

DUAL ADDICTION

Like other addictions, food is a method of coping. "The fat person," according to Stanton Peele in The Meaning of Addiction, "sustains his emotional balance by overeating, perhaps as a way to reduce anxiety or as a sign that somebody—if only himself—cares for him."

Some of the people who come to my seminar have been using cigarettes to control a compulsive desire to eat. They use both food and cigarettes to handle feelings. When they see that they're gaining too much weight, they cut back on food and increase their smoking. When too much smoking makes them sick, they smoke less and consume more food, maintaining a precarious balance.

They don't realize they're dealing with two addictions. And when they try to stop smoking, all their troubling emotions must now be managed with food alone, upsetting the balance. And they rapidly put on weight.

People who depend heavily on food and cigarettes to suppress feelings of loneliness, anger, depression and low self-esteem probably need counseling or group therapy along with their stop-smoking program. Without help, they'll only trade one addiction for the other, gaining more weight and feeling even more unhappy before going back to smoking.

WHAT FEELINGS ARE ABOUT

If we want to be successful at staying off smoking, we need to learn to accept our emotions and the temporary distress that often accompanies them. Our feelings have a purpose. They're not there to torment us, but to tell us something. Feelings of boredom, anger, and depression are signs that something is wrong and that we need to make a change.

Rather than numbing ourselves with food and drugs, we can use our newfound sensitivity to guide us in our search for a more exciting and fulfilling life.

At her wit's end, Vera crept out into the night, armed with a bent-out coat hanger, and fished around in the dumpster till she somehow snagged her cigarettes and pulled them out.

CHAPTER
8

WHEN YOUR METHOD DEFEATS YOU

Willpower, Quick Fixes, Aversion, Distraction: Trivial Attempts to Solve a Serious Problem

"Do you see why all your old methods didn't work for you?" I ask my class. "I'm sure you do, Stephanie. You tried every program that came around."

"It's really clear to me now," she says. "Most of the things I tried were the opposite of what I'm learning now. Just keeping my cigarettes with me has been a big help. I used to throw them away, thinking that if I put enough distance between me and them, I could resist."

"But the question is," I ask, "how far away is far enough? You can get away from the cigarettes, at least for a while, but you won't get away from your desire to smoke, will you?"

Bill tells about the time he stopped smoking on a camping trip with his father. It was an outing he'd been trying to arrange for several years.

"It seemed like the perfect opportunity to stop smoking—far off in the wilderness with no temptations. Besides, my Dad always used to give me a hard time about smoking. He'd say, 'For a smart guy, you can sure do some stupid things.' So I thought he'd be real happy that I'd stopped.

"But it was a long drive, and we got to talking about politics, something we've always tended to argue about. It was then that I really wished I had a cigarette, but I wasn't about to stop and buy any. By the time we hiked into the woods and finished setting up camp, everything was starting to get on my nerves."

"What a time to go into drug withdrawal, Bill," I say. "Was there any way you could have gotten a cigarette out there?"

"No, there wasn't anyone else around for miles, but I kept hoping somebody might show up. I was so obsessed with getting my hands on a cigarette, I went through all my gear and finally started poking around the campfire, thinking maybe somebody might have put out a cigarette under one of the rocks. By the next morning I was so on edge I had to tell my Dad that I thought I was getting the flu, so we could get out of there. What a way to ruin a great trip."

Sadly, out of sight doesn't mean out of mind. On the contrary, the more addicts feel cut off from their drugs, the more desperate they are to get them. Whatever fantastic plots they'll devise to flee cigarettes, they'll dream up equally creative ones to get them back.

Vera talks about finally getting so disgusted with her shortness of breath and her heart palpitations that she tossed her last pack of cigarettes into a huge dumpster outside her apartment. "I just told myself, 'That's it, I quit!' and stomped back into my building," she says.

By midnight she had turned her entire apartment upside down—searched through every coat pocket, cupboard and drawer. Finally, at her wits' end, she crept out into the night in her slippers and robe, armed with a bent-out coat hanger. Climbing up onto the narrow edge of the dumpster, she fished around with the hanger till she somehow snagged her cigarettes and pulled them out.

"I felt like an idiot lighting up out there on the street," she said, "but what a relief!"

Both Bill and Vera were trying to get away from the temptation to smoke. They started out doing something good for themselves, but it backfired. When they threw their cigarettes away, they threw themselves into deprivation.

WILLING IT AWAY

At one time or another, almost everyone who smokes tries to stop through sheer willpower. You wake up one morning and make a firm and determined decision: "I will not smoke anymore." You take yourself in hand and simply go for it.

You may be successful for days or even months. But willpower can work both ways. If you make up your mind to stop smoking, you can just as easily—in a moment of anger, rebellion or despair—make up your mind to smoke.

I will not smoke becomes *I will do as I please.*

Instead of relying on willpower, you need to focus on what you want for yourself. Be ready to make a new choice every time you have a new desire to smoke.

I SHOULD BE ABLE TO STOP ON MY OWN

If you think it's only a matter of willpower, you won't see why you should need any help.

The first night that Mike came to class, I could guess what his problem was, just from the way he sat back in his chair, his arms crossed, with a distant I'm-above-all-this look in his eyes. He came to my seminar because he couldn't stop smoking, but he didn't think he should have to be there. He should have been able to stop on his own. He knew other people who had white-knuckled their way through it and hadn't touched one since. Why couldn't he?

After all, he was sure that stopping smoking was simply a matter of willpower, yet somehow his willpower wouldn't take hold.

Except for medical problems, many of us are uncomfortable asking for help. We feel foolish and ashamed that something as simple as stopping smoking is beyond us.

Men, especially, feel humiliated because they've been beaten by something they believe they ought to be able to control. It's a blow to their ego to admit defeat. Some of them would rather die than admit they lack the strength of character or moral backbone to conquer this "minor vice."

In his 'remember' letter, written during the seminar, Mike said: "I've always been a take-charge type in almost any situation. It's hard to face the fact that I need help with this problem, but I do. Since I finally can admit this, maybe I'll be able to move on to bigger and better things. I owe it to myself."

When a problem is so critically underestimated by society, we naturally feel embarrassed when we can't handle it on our own. And seeing other people "beat the habit" confirms our sense of inferiority.

This has been true of alcoholism and other addictions in the past. Fortunately, attitudes have been changing. When celebrities and well-known public figures reveal their drug dependencies and check themselves in for help, everyone benefits.

Until recently, smoking has not been categorized as a drug addiction, though it kills some 410,000 smokers every year in the U.S. alone. (Another 50,000 people die from causes related to secondary smoke, according to the U.S. Centers for Disease Control). And as long as it's seen as a problem of behavior—a bad habit or a vice—we won't feel justified asking for assistance. When we do understand the nature of the problem and get effective help, we can soon put an end to years of needless suffering.

TURNING OFF

When you're so physically ill that cigarettes become repulsive, stopping is effortless. If something is nauseating, it's not hard to resist. Using this logic, one national stop-smoking program bases its method on aversion therapy. Attaching electrodes to their clients and applying mild electric shock, along with goading them to smoke until they're sick, they create an artificially induced aversion to smoking.

This method is not only demeaning, it's undependable. When the aversion wears off, the desire tends to come back. As any smoker knows, you can be sick and disgusted with smoking one day, and be just as attracted to it a few days later. Why? The body gets well. And if you relapse, are you likely to rush back for another round of this torture—even though the second time through may be free?

Given the right education and support, human beings can think for themselves and make hard choices for themselves.

SCARE TACTICS

Methods that subject smokers to films of diseased bodies, gruesome cancers and surgery also fail to achieve their goals. If you're a long-time smoker who has to have 30 to 40 cigarettes a day, you're probably already scared, if not terrified. But you smoke *in spite of* your fear. In fact, isn't it true that when you're anxious and upset you want to smoke more?

Stephanie told me she became so tense during one of these scare sessions that all she could think about was getting out of there to have a cigarette.

Sometimes scare tactics can work, but too often the effect is temporary. That's what happened to Bill. He spent a lot of time at the hospital when his uncle was dying of throat cancer. It scared him enough that he decided to stop. But six months later, at a class reunion, it all seemed unreal. His fear had long vanished and he reached for a cigarette with hardly a second thought.

FIGHTING THE LITTLE DEVIL

Keeping the devil at bay figures prominently in the method advocated by at least one national stop-smoking program. They tell you to avoid temptations: coffee, tea, alcohol and spicy foods. "Milk or buttermilk is the beverage now. For a hot beverage, use a cereal drink." And since alcohol strikes at reason, judgment and willpower, "confer upon it the dubious honor of being labeled Personal Enemy Number One."

When an irresistible urge comes over you, "Get a drink of water, start deep breathing, and ask for divine aid."

But all too often, the devil wins—since the smoker isn't prepared to talk back effectively to these junkie "temptations."

PEER PRESSURE

Some strategies emphasize peer pressure. If you can't stop on your own, tell all your friends. Set a date and announce the decision to your family and the people at work.

The fateful day arrives, and you get support, encouragement, and premature congratulations, but it isn't long before you begin to feel the pressure of all the attention. Everyone is watching you, wondering if you'll make it this time.

You'll get constant inquiries about how it's going, as well as plenty of other comments: "You're not cheating, are you?"; "Have you blown it yet?"; or, "We *know* you're going to make it this time."

You begin worrying about losing face, about what will happen if you don't succeed. Letting other people down becomes a heavy burden. Now, not only do you have drug withdrawal to contend with, but the fear of looking like a failure, as well. With all these people on your back, you break down and smoke "just one little cigarette"—to relieve the pressure of trying to stay off smoking.

PROVING A POINT

Stopping smoking to prove you can do it usually proves only that. Unhappy with the notion that you're controlled by something, you'll be satisfied once you've proved that you *can* stop. Then when the going gets rough, it's easy to think, "I can always stop again"—which you'll soon find isn't quite that easy to do.

BECAUSE YOU SHOULD

With the medical dangers better known, and the social pressure to stop smoking growing stronger all the time, you may try to stop simply because you feel you should. But it isn't enough to think you *should* improve, you have to really want to change for your own good reasons.

If you're not concerned for your own happiness and health, you won't be prepared to make the necessary effort.

SAVE ME FROM MYSELF

Recruiting friends to help police your efforts is another tactic. In effect, you're saying: "I can't do it. I'm out of control. I can't resist, so you stop me. If I ask you for a cigarette, don't give me one and don't let me take one from somebody else, either." But no one can keep determined addicts from their cigarettes. Soon the "police officer" is resented—then avoided.

Jackie tried this. "I got my husband to promise that he would not let me smoke, to take the cigarette out of my mouth if he had to. Of course, when I began sneaking cigarettes, I had to avoid him more and more. I was always coming home late or running over to my girlfriend's house to smoke. Finally, he started suspecting I was having an affair, and I had to break down and admit the truth."

"You *were* having an affair, weren't you?" I told her. "With your cigarettes."

PUTTING MONEY ON IT

You won't smoke if you bet a lot of money on it, many people believe, trying to lock themselves into their commitment. But people throw their money away all the time. If you can't stop smoking to keep your health, how can you expect to stop merely to keep a little money?

And who will make sure you hold to the bargain? As the going gets rough, it's easy to rationalize, "Who's going to know?"

When the truth finally does come out—well, it's only money.

KEEP MOVING

The most common advice you'll hear is: keep busy.

But if you jog around the block or wear yourself out cleaning the house or mowing the lawn, sooner or later you've got to take a break. What will be waiting for you then?

Inevitably, the desire to smoke.

What then? More jogging, more cleaning? How about brushing your teeth or going to a movie? Then what? Would anyone offer such counsel to an alcoholic or cocaine addict?

Even though the Surgeon General has called nicotine the most widespread example of drug dependency in the country, the major health organizations continue to treat smoking as a bad habit. And their anti-smoking campaigns give out reams of naive advice.

ACUPUNCTURE & HYPNOSIS

Clinics offering acupuncture and hypnosis as a way to stop smoking often claim extremely high success rates. These claims have never been backed up by objective research. And what do they mean by success? Stopping smoking for one week, one month?

Hypnosis or acupuncture may temporarily relieve your craving to smoke and help you get through withdrawal, but these techniques don't give you the tools to stay off smoking over the long haul. Of course the thought of having your desire for cigarettes taken away, as if by magic, is very appealing. But even if your urge to smoke temporarily disappears, you can be certain that sooner or later it's going to come back. If you don't know how to handle it when it does, you're going to be in trouble.

There's no substitute for the time and effort it takes to work through your drug addiction.

NICOTINE PATCH

In 1992, the nicotine patch was introduced with great fanfare and a multi-million dollar advertising campaign, depicting smokers radiating confidence and control. Sales of the patch, which peaked at one million orders a month, quickly and dramatically plummeted, prompting one industry analyst to comment that he had never seen such a downturn in a new product.

Obviously, the patch wasn't living up to its promise. What happened?

Enticed by an image of a cure that seemed effortless and painless, smokers ignored the fine print on the ads--which recommended that the patch be used as part of a comprehensive treatment program. Success rates for the patch, when used on its own, were only about eight per cent, according to a 1993 report from the U.S. Centers for Disease Control.

Designed to deliver nicotine through the skin in diminishing doses over several weeks, the patch can lesson symptoms of withdrawal. As such, it can be helpful, especially for long-time smokers, as part of a treatment program.

But the patch usually won't eliminate your desire to smoke. Though you're still getting a steady dose of nicotine, it just isn't the same as smoking. The patch doesn't deliver nicotine with the same punch you're used to. To compensate, some people sneak cigarettes while using the patch, a dangerous practice that can trigger a heart attack.

Moreover, the patch doesn't teach you how to cope with life without your cigarettes. Eventually, you will have to learn how to make the psychological and emotional break with a drug you're deeply attached to. Without this knowledge, all it takes is an upset or a little stress to trigger a relapse.

CUTTING DOWN

Cutting down is one of the least effective ways to overcome nicotine dependency. Does anyone recommend that alcoholics or cocaine addicts cut down—rather than just stop?

This method aims primarily to diminish the physical symptoms of withdrawal. But in my fifteen years of teaching stop-smoking classes, people seldom report physical reactions more serious than mild headaches, sweaty palms, blurry vision, or insomnia—reactions which almost always disappear after a few days. People who give up coffee often complain of worse side-effects.

Rather than being helpful, tapering off can set you up for failure. When you cut down on cigarettes, you put yourself in limbo. Neither a regular smoker nor yet a successful ex-smoker, you prolong the phase of withdrawal from a few days to a week or more. Any minor crisis can trigger your desire for a cigarette, and since you're still smoking, it's easy to justify having a few more than you intended. Once you lose your momentum, you tend to smoke more—not less, and you're soon back to full-time smoking.

NICOTINE GUM

Like using the patch, chewing nicotine gum can help you avoid some of the symptoms of withdrawal for a while, and it's certainly better than smoking. But if you want to kick your drug addiction, you're going to have to face withdrawal at some point. If you keep chewing the gum indefinitely, you've just traded one way to get your drug for another.

HELPFUL HINTS FROM THE AUTHORITIES

Imagine a veteran smoker, in the throes of nicotine withdrawal, trying to fend off intense cravings with the following helpful hints, typical examples taken from current stop-smoking literature.

"Immediately after quitting . . . strike up a conversation with someone instead of a match."

That's cute, but as a fledgling ex-smoker, do you think you'd be in the mood for small talk? And if you're like most smokers, you use cigarettes to calm your nerves when "striking up" a conversation.

"Miss the sensation of something in your hand? Play with a pencil, a paper clip, a marble."

You can fiddle with anything you like, but your hands aren't the problem. What you miss is the relief you get from using your drug.

"If you always smoke while driving, take public transportation for a while."

Obviously, whoever dreamed this up has never waited for a bus.

"Don't sit in your favorite chair, or have a cocktail before dinner. Avoid friends who smoke and favorite meeting places. Spend time in the library, museums and churches."

To follow this advice, you'll have to rearrange your whole life. Not only do you give up smoking, you give up almost everything else you enjoy. "Life just didn't seem fun anymore," says Vera, who tried this tactic once. "I had to change my whole life and felt left out of everything."

To stop smoking successfully, learn to work through the desire to smoke whenever it occurs, rather than disrupt your whole life trying to avoid it. In fact, the more often you are exposed to activities you associate with smoking, the sooner they will stop triggering desires to smoke. But use common sense. Since alcohol affects judgement, stay away from wild partying or excessive drinking during the first few weeks of recovery.

"Take a shower when you feel the urge to smoke."

During the first two days of withdrawal, there are times when the desire to smoke is almost constant. That's a lot of time in the shower.

While the suggestions above range from the frivolous to the preposterous, others may be good for your health, but they are inadequate for treating drug withdrawal:

"Get plenty of rest."

It's a good idea to try, but one of the most frustrating reactions to stopping smoking can be insomnia. Fortunately, it's only temporary.

"Swim, jog, play tennis or handball."

Sports and exercise are a wonderful discovery for ex-smokers who've been deprived of the joy of being physically active. But if you hate sports, and in fact resist all physical effort, this admonition will hardly cheer you on.

"Drink a lot of water in order to flush the nicotine out of your system."

Drinking a lot of water may be a healthy thing to do. But flushing nicotine out of your system quickly isn't going to make you crave it any less. It's not the nicotine *in* you that makes you want to smoke, it's the *lack* of nicotine. If it leaves your system naturally, you have more time to adjust, to make the difficult transition from smoker to ex-smoker.

DANCING AROUND THE PROBLEM

Programs that offer little but hearty advice and well-meaning tips may help someone who's disciplined or only mildly attached to cigarettes. But when they're aimed at compulsive smokers, they're virtually useless.

Substitution, distraction, avoidance and willpower—these strategies stem from a philosophy that treats smoking as a bad habit or vice, a question of character or morals. They encourage you to dance around the issue, distracting you from the underlying truth: you are addicted to a subtle but powerful drug. To break free, you'll have to confront your addiction and work your way through it.

If you try to avoid this reality, you'll be building your house of straw or twigs like the first or second little pig. You'll be defenseless against the big bad wolf—the sudden, sharp craving for a cigarette. He'll huff and he'll puff and he'll blow your house in.

"I've been missing smoking lately," says George, during a follow-up call. "But maybe that's not surprising—since I smoked for 50 years."

CHAPTER

9

DOWN THE LINE

AVOIDING RELAPSE:
PREPARING FOR HIGH RISK
SITUATIONS & EMOTIONS

Of all the traditional New Year's resolutions, quitting smoking is one of the first to go up in smoke. It's considered an event: you make the decision, throw away your cigarettes and expect it to be smooth sailing from then on.

In reality, becoming an ex-smoker is not an event, but a process, a daily set of challenges. Not only are you addicted to nicotine, your drug is aggressively marketed; and the sight and smell of cigarettes continually tantalizes you, reawakening your desire to smoke. So you need to be prepared to deal with your desires every time they are triggered.

After they stop smoking, my clients have no idea exactly what kinds of pitfalls await them over the next six months. Each will have individual challenges to face, situations and emotions they'll confront for the first time as ex-smokers. And each will need individual help to guide them through these high risk situations.

FEELINGS OF LOSS

When I call George, six weeks after he's stopped, he sounds a bit cantankerous.

"What's going on, George? You've been doing so well."

"I guess I'm feeling a little rebellious out here. I'm used to getting my way. I was just thinking how nice it would be to smoke a cigarette."

"Sure you'd like to smoke a cigarette. But do you think one cigarette would do it for you, George? Has it ever before?"

"Nope."

"All right. Then exactly what are you rebelling against? That you can't smoke just one? If you could have it your way, is that how you'd have it?"

"Probably."

"I've got some bad news for you, George. If you try to smoke just one, you'll be stuck with smoking them all."

"Well, I just feel like I'm missing something."

"How long did you smoke? Fifty years?"

"About that."

"Then of course you're going to miss smoking. It's normal. But do you miss not being able to breathe, George? You've got a hard choice to make. You can give in to this temporary rebellious feeling or you can take care of your breathing. You have a feeling of loss, and you have lost something. But it isn't cigarettes. They're a five-minute drive away. You can still smoke. What you have lost is the ability to smoke without discomfort. And you'll never get that back. The party's over."

"I guess I realize that."

"Sure you do, when you take time out to think about it. You miss smoking now, George, but it's only been six weeks. As time passes, you'll be missing it less and less, until you'll hardly ever think about it. Stick with it, you'll see."

ON THE ROAD

Four months into her recovery from nicotine addiction, Vera wins a trip to Reno playing bingo at her Church.

"Oh, I'm so excited about it," she tells me, when I check in with her on a routine follow-up call. "This is the first time I've felt like doing anything like this in years. And if I were still smoking, I just know I wouldn't be up to it."

"Congratulations, Vera," I say, though I'm thinking to myself, 'Why does it have to be Reno?' "Enjoy your new-found energy. But let me warn you about getting caught up in the atmosphere there, all that exciting nightlife. It's easy to forget how important not smoking is to you."

"Oh, I don't think I'll have any problems."

"That's what worries me, Vera. I know you've been doing very well here at home, but going on a trip is one of the riskiest situations you can put yourself in as a fledgling ex-smoker. I've worked with many people who loved not smoking, but in a place like Reno they've been willing to gamble on a long shot—that they could get away with just one. Believe me, you're going to be tempted."

"You mean I shouldn't go?"

"To tell you the truth, unless you're going to take this risk seriously and prepare for it, you should stay home."

"Well, I don't want to stay home, so what should I do?"

"First of all, expect to have strong desires to smoke in Reno. Don't go there hoping it won't bother you. Just imagine yourself in one of those nightclubs. There you are being served drinks, with music and dancing and lots of people smoking around you. When the compulsion to smoke descends over your brain and the junkie thinking starts working on you, you're going to need something *real* to grab onto. So take your notes from class with you."

"Oh, yes, I still keep them in my purse."

"Good. Now you won't have any trouble if you do this: take time out. When you feel the tension building, don't try to bully your way through it and hope for the best. Go off by yourself in a quiet place, the lobby or outside, and review the steps you learned to handle the desire to smoke. All it takes is three minutes. You'll come to your senses and when you get back to your table, you'll be able to see all that compulsive behavior around you and feel proud that you haven't been swept away by it. Do you understand how that will help you?"

"Yes."

"Okay. Here's something else you need to do. Before you leave for the airport, spend a few minutes reviewing the steps you learned to handle the desire to smoke. Do it again when you get to your hotel, and whenever a new desire comes up. Now tell me how you're going to think it through."

"First of all, I'm going to remind myself that I can smoke because I'm not deprived. But I can't smoke just one and not go back to smoking them all, so I have a choice to make. Then I choose to be temporarily uncomfortable rather than give in because I want a healthy heart, no more fear and worry, better breathing, and more energy, so I can keep on enjoying myself."

"Great, Vera. One last point. With each new situation on your trip, stop and ask yourself this: 'Do I intend to get through this as an ex-smoker?' If so, give five good reasons why."

Intentions are important. You never know what you're going to do until you're actually doing it. But you can know what you intend to do, and that intention will help give you the strength to get through it.

If you don't think traveling is risky for ex-smokers, consider all the things you'll be doing that you normally associate with smoking: socializing, celebrating, drinking, and meeting old friends who still smoke. And what if you lose your luggage, get lost, miss your connections or get stuck in an airport overnight? Often, just being out of your own environment makes you feel disoriented and anxious.

But none of this is too hard to get through if you're prepared for it and willing to take the time to think.

TROUBLE AT WORK

When Stephanie calls me two months after stopping smoking, she's almost in tears. I've been working with her for a few weeks on the problems she's been having at her job. As long as she was smoking, she was able to tolerate her editor's blatant arrogance and sexism. She watched as junior reporters, all men, were promoted ahead of competent women who'd been there just as long.

Lately, instead of going out to the lounge to numb her feelings with cigarettes when she got angry, she'd spoken out and had already written up her complaints and submitted them to the managing editor.

"Guess what?" she says. "I'm not at the newspaper anymore."

"Did you finally decide to quit, Stephanie?"

"Not exactly. I got fired today. Even though I've been fed up with this job, now that I've lost it, I'm really scared. I have house payments and I just bought a new car. And I must confess, I've really been wanting to smoke today."

"Well, that isn't surprising, is it? I'm glad you called. Let's work through it."

It takes several phone conversations during the next week before Stephanie comes to terms with the upheaval in her life and realizes that she doesn't have to smoke herself to death over it.

She has just become comfortable again as an ex-smoker, when she calls to announce that she's been hired as a reporter for a television station. "I'm so thrilled about it. It's something I've always wanted to do. But I'm really nervous, and my desires to smoke are really strong again. And this job seems more important than not smoking."

"You seem to think you have to choose one or the other, your job or your health. But that's not true. And to act like your job is a higher priority than your breathing is crazy. They're both important and you can have both. But you'd better correct that junkie thinking before you make it okay to use drugs to handle your feelings. You're nervous now, but after a couple weeks on the job you'll be perfectly comfortable. And you won't be stuck with smoking."

Anyone can have problems at work. But remember that addicts want to use drugs to solve their problems or numb the pain of having them. So if you're not careful, you'll use any problem as an excuse to give in and smoke.

ON THE ROCKS

"How's it going, Mike?" I ask on a follow-up call three months after the seminar.

"Fine," he says, in his usual noncommittal style.

"Really? No problems at all?"

"Except that I had a hell of a fight with my wife the other night."

"Did it bring out any desires to smoke."

"You bet. In fact, somebody left cigarettes at our house and I had one out of the pack before I stopped myself."

"How did you get yourself to go that far, Mike?"

"I was so mad, I just thought, I'll show her."

"You mean, 'Now look what you made me do'?"

"Something like that."

"How'd you stop yourself."

"I remembered that I'm the one who'd have to go on smoking and pay the price. I didn't stop smoking to prove anything to her, so why should I have to start again to show *her* something? I stopped for my own good reasons."

"Good thinking, Mike. I'm glad you're clear about that. How are you and your wife doing now?"

"Oh, we got over it. It wasn't that big a deal."

"It wasn't worth relapsing over, was it?"

"Heck, no," Mike agrees.

GLAMORIZING

About the same time as my conversation with Mike, Lisa tells me about going out for a drink with a friend of hers who still smokes.

"We were just sitting and talking at an outdoor cafe and I sure wasn't having any cravings to smoke. But as I was watching the people around me who were smoking, my mind started toying with the idea. They looked so cool, like smoking wasn't hurting them at all. And you're not going to believe this, but I started thinking, 'Maybe it wasn't really hurting me either'."

"Haven't you done that in the past?" I ask.

"Yes. But this time I caught myself, 'What am I doing?' At that moment I remembered something that happened a couple of months before I started your class. I had a sudden vivid memory of myself sitting in my bathtub sobbing because I had been unable to stop smoking. I'd come back from a trip to California, where my grandmother was in the hospital dying of emphysema. I realized I already had some of the same symptoms she had, and I was scared. So I tried to stop smoking right away as soon as I got home. I really freaked out when I couldn't.

"And here I was, six months later, thinking about how glamorous smoking was. I told myself, 'Get real! I'm having an urge to smoke and I can smoke, just like these people are. But I'd have to smoke just like they do—compulsively and with all the consequences.'"

"And how did you feel when you got through it?"

"Wonderful. I felt so relieved that I didn't act on such a silly impulse like that."

Even happy, comfortable ex-smokers need to be careful about

relapse. It's easy to become overconfident, to feel you've got it made, to forget what prompted you to stop. The farther you get from withdrawal, the easier it is to glamorize smoking.

HIGH RISK

Without adequate understanding and effective skills, you can hardly expect to get through all the situations that pose a high risk for drug addicts trying to stay off their drug.

These include: emotional upsets of all kinds, such as tension at work, getting fired, breaking up a relationship, arguments, accidents and depression; illness, pain and fatigue; socializing, drinking, celebrating, meeting old friends who smoke; and traveling and vacations.

Here's how to prepare yourself to get through any high-risk situation:

(1) Expect to have strong desires for a cigarette and treat these urges the first time they come up. Don't try to ignore them, or they'll just keep coming back.

(2) Take time out. When you have a craving to smoke, go off by yourself for a few minutes and work through the five steps for handling your addiction.

(3) Ask yourself: "Do I intend to get through this situation as an ex-smoker? If so, why?" Then name three of your benefits.

Besides events and situations, certain feelings also pose a high risk for ex-smokers. After all, people use drugs to alter or numb their feelings.

High-risk feelings include: anger, frustration, self-pity, depression, excitement, grief, disappointment, fatigue, boredom, fear, hopelessness, loss, anxiety and worry.

These feelings tend to trigger junkie thinking, such as: "Who cares?" "What difference does it make, anyway?" "Why bother?" "I really need one now."

Think ahead. Which emotions will be the most dangerous for you?

Is anger one of them? If so, remember that you're going to be either an angry smoker or an angry ex-smoker. Smoking doesn't take away your anger.

The same goes for other feelings: depression, boredom, fear. Smoking doesn't take them away, it only takes the edge off. And you pay a very high price. Feelings are temporary and always changing; smoking goes on and on.

Think before you act and talk back to your junkie thinking. If you do, none of these feelings can force you back to smoking.

STOPPING FOR THE WRONG REASONS

Staying off smoking is tricky anyway, but if you stop for the wrong reasons, it's nearly impossible.

There's one point I always make clear at the beginning of a new class: "If you're here to please somebody else, or to get someone off your back, don't bother. Sure, you can stop, but you won't stay off."

To be successful, you've got to stop smoking for your own sake, not for the sake of your husband, wife, friend, or child. If you depend on pressure or support from other people, it's likely to backfire after a while.

It won't be long before your heroic efforts are taken for granted; the praise and encouragement lavished on you at first soon subsides. And when the person you're trying to please forgets to pick you up after work or leaves a big mess in the kitchen for you to clean up, you'll be angry, disappointed and ready to smoke: "After what I gave up for you! This is how you treat me?

Okay, watch this."

Or you may get in a heated argument and end it by grabbing a cigarette: "Well, I hope you're happy now, dear. Look what you made me do."

PMS

For women with premenstrual syndrome, stopping smoking during this stressful time can intensify the discomfort of withdrawal, making it seem like agony. They may even attribute all the stress and emotional turmoil they're feeling to the fact that they're not smoking and decide that stopping is just too difficult for them.

If you're a woman who suffers from PMS, it's best not to stop smoking during this critical time, in order to avoid the added stress and greater chance of failure.

UNDERESTIMATING THE PROBLEM

Even with the best motivation—a nagging chest pain or the threat of a heart attack—you're still likely to relapse if you underestimate the problem of addiction. You may know that smoking is killing you, but you also need to understand how it's *running* you. You need to learn how to control a compulsion, a drug addiction.

As amazing as it may seem, even if you've been smoking a pack and a half a day for 30 years and feel totally dependent on cigarettes to make your life work—smoke everywhere you go and with everything you do, before closing your eyes at night and as you're opening them in the morning—even then you probably expect a quick and easy solution. You want to stop smoking effortlessly. You want magic.

The newspapers are full of ads from the would-be magicians, promising cures with "No Effort," "No Withdrawal," and "No Discomfort." Many claims are exaggerated, with unsubstantiated 80 to 90 percent success rates.

So you go to the hypnotist, the acupuncturist, the aversion therapist with the attitude: "Okay, doc, take it away! Get rid of it for me." If the treatment does work, you walk out thinking it's all over, that you'll never have to worry about it again.

And maybe you won't smoke for a while, but chances are you've set yourself up for a relapse, because you're not prepared to deal with your emotional life as an ex-smoker. Your attachment to cigarettes, though latent or repressed, is still very strong; you've formed a bond with them and reinforced it for many years. And you won't have learned what to do with the inevitable urges to smoke or how to fight back against junkie thinking.

TIME DOESN'T CURE

If you believe that time cures addiction, sooner or later you'll decide it's safe to have just one.

Though the desire goes latent, often for long stretches of time, that doesn't mean you're cured.

Though smoking is not as socially acceptable as it used to be, you'll still find yourself in situations that will trigger your desire to smoke, no matter how much time has elapsed. If you don't know how to defend yourself and don't have the skills to fight off the subtle persuasion around you every day, you're going to succumb to your desire, sooner or later.

And armies of highly-paid professionals are working full time to get you back in line at the corner convenience store.

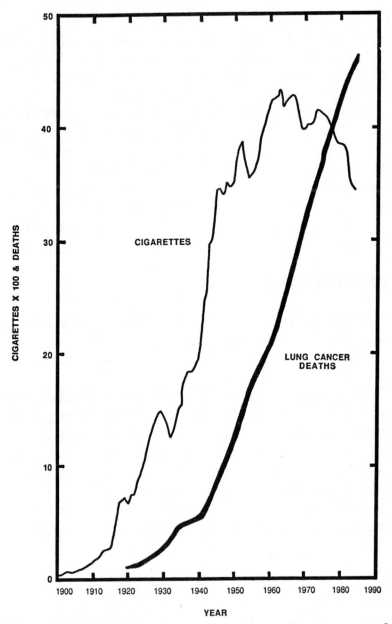

Before 1920 lung cancer was extremely rare, and physicians were unlikely to see it in a lifetime of practice. After 1920 incidents of lung cancer escalated, following a 20 year boom in cigarette sales.

CHAPTER
10

WHO WANTS YOU TO SMOKE?

Big-Time Drug Pushers, Profits & You

If you want to become very wealthy, one of the most lucrative businesses you can get into is the drug business: cocaine, marijuana, heroin—or tobacco.

Not only are drugs cheap and easy to produce, but your clients become dependent on your product indefinitely, sometimes for life. And they'll buy from you regardless of the price. Unlike other enterprises, the drug business doesn't suffer from downturns in the economy. When times are bad, people tend to rely even more on drugs, seeking relief from stress and anxiety.

Cocaine dealers gladly give away the first line to a potential user. "Why don't you just try some?" they offer. "Here. It's free."

Like these drug pushers, cigarette companies make it as easy as possible for you to get started. They even give away free cigarettes on the street corner. A recent ad in Parade magazine, pushing smokeless tobacco, says: "We are so sure you're going to like it, we're going to give you a can—free!"

Such a ploy can even snare people years after they've stopped smoking. "One day when I was downtown I was given a free pack of cigarettes, so I decided to give them to a friend," Bill told me, about six months after taking my seminar. "That night I had an argument with my girlfriend, and after I left her house I discovered the cigarettes in my jacket. I already had the wrapper off and a cigarette out of the pack, when suddenly I realized how I'd been set up and I threw them away in disgust."

What happens when an ex-smoker, still in the throes of withdrawal, is offered a free mini-pack of his favorite brand? It's easy to rationalize, "Well, I'll just smoke these since they're free—and there's only a few, anyway."

Unlike illegal drug pushers, tobacco promoters have free rein to make it as difficult as possible for you to stop. To avoid temptation, a heroin addict can move out of the dangerous environment. But it's nearly impossible to escape the long reach of R.J Reynolds and Phillip Morris. Their ads in newspapers and magazines extend right into your living room.

Ex-smokers really want to be like non-smokers: people who never think about smoking. Too bad. A multi-billion dollar industry isn't about to let that happen. Every customer they lose costs them hundreds of dollars each year.

When a heroin addict once asked me why it's easier to stay off heroin than to stop smoking cigarettes, I said: "For one thing, do you have a billboard in your neighborhood showing a handsome fellow with a big smile on his face and a needle hanging out of his arm? Are you exposed to the wonders of heroin on every other page of magazines and newspapers? Are you constantly seeing junkies tying off and shooting up at every bus stop and restaurant?"

KEEPING YOU ON THE HOOK

It takes several months to make the transition from smoker to comfortable ex-smoker. And during those months, what do you think you're being hit with every day?

Ads that imply: if you were smoking right now, you'd be more ambitious, more stimulated to succeed. If you were smoking right now, you'd be slimmer and more attractive. If you were smoking right now, you'd be having more fun, you'd be more outgoing, sexier, more sophisticated.

And if you don't think it through carefully, chances are you'll believe them. No matter how much suffering may have caused you to stop smoking, you still have an image of yourself as a happy, carefree, confident smoker.

Cigarette advertising does everything possible to create a reality that simply doesn't exist.

Out of the 20 to 30 cigarettes you smoke every day, how many do you smoke for that roasted tobacco flavor? Yet in withdrawal, you're convinced by the ads that you smoke for "TASTE," for FLAVOR YOU CAN FEEL ALL THE WAY DOWN TO YOUR TOES.

You're not a failure when you can't stop smoking. The ad says "VICTORY!"

You're not feeling tired and dragged out, you're feeling "ALIVE WITH PLEASURE!"

It's not that same old boring cigarette, "IT'S A WHOLE NEW WORLD."

A DRUG BY ANY OTHER NAME

Cigarette companies are all selling the same thing: a drug experience. It's the same drug—nicotine—in every brand of cigarettes.

But cigarettes aren't marketed as a drug. Instead, the focus is on the *effect we seek* from the drug. And different brands promote different effects.

When you stop smoking and intend to stay off the hook, you'd better realize how determined the cigarette companies are to lure you back. They are masters of persuasion, and their ads are created by experts in social psychology. Supposedly trained to understand and help people, these professionals use their knowledge to play upon human fears, anxieties and inadequacies to induce us to consume in predictable ways.

When I ask my clients to examine cigarette ads closely, they're surprised at the subtlety and sophistication of this bait.

I hold up magazine ads for Vantage and Merit, one in each hand. "What's going on here? All you've got are pictures of boats. Nobody's even smoking."

"They're selling the two opposite effects of nicotine, aren't they?" Stephanie says. "The speed boat in the Vantage ad is pushing excitement, and the one for Merit is totally mellow and relaxed—this beautiful sailboat gliding over the sea."

"Yes," I say. "And they've got something in common, too. They're both saying, 'Get away from the rat race. Get away from it all.' Isn't that the feeling you want when you take a cigarette break?"

Next I show them an ad for Triumph, an expensive, two-page spread with close-ups of four smokers grouped together and grinning fiercely, fists raised and clenched, a cigarette protruding up from between the first and middle finger. Splashed across one of the pages is the word UMPH!

"What do you think they're trying to get across with this ad?" I ask.

"I don't know, but for smokers, they sure have white teeth," Vera says.

"Yes, they do. Anything else?"

"I see a lot of energy," Bill says, "They look really pumped up."

"That's right. So this ad isn't focusing on the tranquilizing effect of nicotine, it's emphasizing how stimulating it is. But what else is going on here?"

I wait while they take a closer look.

Mike says, "Looks to me like they're flipping us all the bird."

Lisa gasps. "That's exactly what they're doing."

"Isn't it amazing what you see when you look closely? Here are four defiant, rebellious smokers. They're telling the anti-smokers and the American Cancer Society, 'Up yours, baby! Nobody's going to tell us what to do. We will *triumph* over lung cancer, heart disease and clean air laws.' This ad assures you that you're not alone, you're in good company."

"My goodness," Vera says, "I can't believe they actually made this ad."

"They did it because they're brilliant at pushing all our unconscious buttons. Look at this ad for Kent," I say, showing another example. "All you see are two shadows of a man and a woman rushing into each other's arms, against a background of a field of flowers. How does this ad sell cigarettes? It links the high of a drug experience to the high of another experience that we all long for—in this case, the excitement of falling in love. In fact, the headline at the top promises that in this pack you're going to find, 'The experience you seek.'"

"Now, do we have any Camel people in here?" I ask.

Norman and George both raise their hands.

"Okay. Let's look at the Camel man. You've seen this dude all over town, haven't you? In this first ad we see him out in a jungle, sitting on the side of his half-finished canoe, taking a cigarette break. Notice the serenity here. This man has nothing but time. In fact, he's got so much time that he's whittling his own wooden canoe out of a big log with that tiny little hatchet."

"He doesn't have any deadlines, does he?" Stephanie says.

"That's for sure. Now here he is again," I say, holding up an ad which shows the Camel man in an exotic foreign setting. "He's relaxing in a busy marketplace in Timbuktu or somewhere, using his drug without a care in the world.

"Here's another one. Now what's he up to? He's out in some craggy cove by the ocean, standing up and maneuvering his rubber raft with a long pole. Nature looks a little aggressive here, but he's okay. No matter how rough the waters get, he stays calm and unruffled. He can handle anything, can't he, as long as he's got a cigarette hanging out of his mouth."

"I've always hated this guy," Jackie says. "He's such a loner, so smug and detached from everything."

"Yes, and look how laid-back he is. He has no relationship problems, family commitments, or responsibilities. Notice how in all these ads he looks independent, completely self-sustained, relaxed, totally in control. Isn't this what you want from smoking? To keep your life under control by taking time out to shoot up your drug, so that you won't be bothered by stresses and strains?"

"He's never with anybody else, is he?" Stephanie asks.

"Well, look at this one," I say, holding up the first half of a two page ad of the Camel man driving along in his jeep. "And guess what? For once he's not alone." I show the second half. "He's got the companionship of his trusty mutt."

Amid the laughter, Vera says, "But I bet that's not his best friend. I bet his cigarettes come first."

I take out another set of ads for Benson and Hedges, showing people together in fancy restaurants, laughing, having intimate cozy conversations.

"You don't like that Camel guy, Jackie," I say. "But there's lots of other brands to choose from, aren't there? I noticed when I crumpled up your pack the other night that you smoked Benson and Hedges. These ads are pushing success, sophistication, togetherness, self-confidence. Wouldn't we all like a little more of that? And look what the ad says: 'For people who like to smoke.' These people aren't addicts; they don't smoke because they have to. They smoke only because they like to."

"I just wish it were true," Jackie says.

"Yes, and the drug peddlers are doing everything in their power to keep you believing the illusion—that cigarettes will make you free and independent, relaxed and comfortable with your friends. They don't want you to face the truth about your addiction. Because if you do face it, you might have to do something about it."

LURING NEW CLIENTS

Every day while some one thousand Americans are dying from smoking-related illnesses, cigarette companies are aggressively luring thousands more onto a path to the same fate.

Just as drug peddlers hang around school yards because they know that kids are most susceptible to their wares, cigarette promoters know their success also depends on young starters.

Of the thousands of clients I've worked with, almost all of them started smoking as teenagers, many before the age of 17 and some as early as age eight or nine.

They didn't know any better. It was such an innocent act, lighting that first cigarette. They were ignorant of the consequences and never dreamed that something so common in everyday life, so widely promoted, could be harmful, even life-threatening. And few ever imagined that they might be smoking for the rest of their lives.

Even though it's illegal to show kids smoking in ads, smart advertisers know how to appeal to youth. In a confidential memo obtained by the Federal Trade Commission, an advertising company advised this strategy: "In your ads, create a situation taken from the day-to-day life of the young smoker, but in an elegant manner. Have the situation touch on the basic symbols of growing up and the maturity process."

An overwhelming percentage of cigarette ads today appear in magazines for young people. And in women's magazines, almost all of the ads are focused on slimness.

Ads for tobacco imply that life with cigarettes is more adventurous, exciting, and fun, and that you're sure to be popular. The enticing images form a backdrop to celebrities who, cigarettes in hand, bring these images to vibrant life.

These rock stars, film and television idols, writers and artists, commandos and cowboys, exude confidence, success, and sexual allure. Hoping to absorb the same qualities, teenagers emulate such figures, imitating their fashions, style, speech—all the subtle or flamboyant posturing of smoking.

Even ads that portray sophisticated, successful adults appeal to youth's desire to feel grown-up and self-confident.

Kids smoke because it's glamorous, sexy, tough, or cool.

And while many schools are beginning to educate children on the dangers of cigarette smoking, too many young people still have no idea of the reality of addiction. How could they? The truth is all but screened out by illusion, so successfully created by the tobacco companies.

HOW IT REALLY IS

But should we believe the manufacturers? Simply take their word for it? Let's hear what some of their most loyal and dedicated customers have to say. They've been using the product for most of their lives; they ought to know.

Here's how some of the people in my class describe smoking in the 'remember' letters I have them write to themselves:

> Dear George,
>
> Remember listening to the rasping sound of your breathing at night, and changing positions in bed so that the sounds will stop. I wanted it to stop because I knew it was my lungs warning me of impending trouble. The choking, the burning mouth, the pain in the chest and shortness of breath made me even more anxious. At that time there was only one question: how long to live?

> Dear Jackie,
>
> Remember the time that you woke up in the middle of the night with your heart "turning over" and questioning whether you would be there to see your

daughter graduate from high school. But you felt
powerless to change and stop smoking. The horrible
taste of cigarettes in your mouth, the persistent
chest pains, the heart palpitations, the cough that
your husband told you went on all night. Remember
how angry you used to be with yourself, how
incompetent you used to feel, how idiotic and
unattractive. How depressed. Think of it, 16 years
that cigarettes have run your life, and interfered
with your development as an emotionally and
physically healthy human being.

Dear Vera,
    You don't want to forget lying on your side in
bed unable to move and barely able to breathe
because of the awful stabbing pain in your chest.
Even after heart surgery, you still went on smoking.
Remember the fear and self-loathing and the endless
disappointment. And remember your emaciated father
on his deathbed the day before he died of lung
cancer, because of smoking.

Dear Norman,
    How can you forget having a lung collapse and the
total terror of crying out for help. Obviously, you
can't hide that away in the dark recesses of your
mind, along with the nights in bed listening to your
lungs wheeze and flutter, wondering if it's already
too late.

This isn't exactly what the ads promise, is it?

SOLD OUT

Cigarette advertising is big business. Cigarettes are one of the
most heavily advertised products in the USA. Cigarette companies have become the heaviest users of newspaper, magazine and
outdoor display advertising, according to the American Medical
Association.

Such massive spending is bound to have an effect on the publishing industry. Until recently, little information on the dangers of smoking reached the public. Even now, most magazines avoid hard-hitting articles on the horrors of nicotine abuse for fear of endangering their major source of revenue.

While allowing that television stations can afford to be "totally truthful as often as they please" about smoking, the editor of Cosmopolitan, Helen Gurley Brown, excuses magazines for their lack of coverage. "Who needs somebody you're paying millions of dollars a year, to come back and bite you on the ankle," she explained in an interview.

Readers Digest, one publication that frequently exposes the ugly realities of nicotine addiction, has chosen to do without tobacco advertising. They estimate it has cost them several hundred million in potential revenue over the past 30 years.

"Until recently," I tell my clients, "you knew smoking was bad for you, but you didn't get much more information than that. You wouldn't often run across articles telling you that the pain in your chest or the fluttering of your heart was caused from smoking. You really didn't know that smoking would probably shorten your life span by about seven years."

Even many years ago, there were brave voices trying to get the message across. But it wasn't until 1964, when the first report on smoking by the U.S. Surgeon General made such an impact on the country, that people started to take notice.

DENYING REALITY

Surgeon General C. Everett Koop's 1988 report leaves absolutely no doubt about the nature of nicotine addiction and the reality of smoking and disease, calling smoking the chief single avoidable cause of death in our society.

Not surprisingly, the tobacco industry blasted the report. "The claims that smokers are 'addicts' defy common sense and contradict the fact that people quit smoking every day," said a spokesman for the Tobacco Institute.

Koop countered by saying that heroin addicts stop every day, as well.

Addiction doesn't mean that you can't stop, it means that once you're hooked, you have to overcome formidable obstacles in order to get free and stay free.

Backing up the industry, Senator Terry Sanford, from the tobacco growing state of North Carolina, said that Koop had mistaken the enemy. "In comparing tobacco—a legitimate and legal substance—to insidious narcotics such as heroin and cocaine, he has directed 'friendly fire' at American farmers and business."

"I haven't mistaken the enemy," Koop retorted. "My enemy kills 350,0000 people a year." By comparison, 6,000 people a year die from opiates like heroin and about 125,000 people a year from alcohol, he said.

Yet the cigarette companies continue to deny the facts, as they have for more than 20 years. Fighting for their survival, they have been doing everything in their power to create clouds of confusion. In the face of overwhelming evidence, these drug peddlers loudly protest "no proof," claiming that irrefutable facts are "controversial" and "a matter of debate." Though spokesmen for the industry endlessly argue the connection between health and cigarettes, the debate is ludicrous if you consider the obvious. When you stop smoking, you stop hurting.

In one of these debates, I would like to ask: If smoking didn't cause my symptoms, where did the chest pain go when I stopped? What happened to my cough, my bronchitis, the heart palpitations? Why is it that every symptom I had when I was smoking disappeared when I stopped?

And I'm not an isolated case. I've seen this happen over and over again with hundreds of my clients.

Further muddying the issue, the tobacco companies shift the argument to distract you from the real problem. Defending themselves in paid editorials by the official-sounding Tobacco Institute, they focus on individual liberty and free choice, taking advantage of fear of government interference in private life.

"What's the point of all this?" Jackie wants to know.

"The point is, if they can maintain a little doubt in your junkie mind about the dangers of smoking, they've got you! And that's what you're looking for, isn't it? Some way to rationalize your behavior in spite of your fear. Then you can tell yourself, maybe the Surgeon General is wrong."

As long as the "debate" churns, the profits keep rolling in. And people continue to pay the deadly price.

PREYING ON THE POOR

How does the tobacco industry get away with it?

Money and power. They've got the money and they buy the influence. All over the world, cigarette companies lobby governments, which collect huge revenues from taxes on tobacco products.

As people in this country get smart and break out of the trap, the tobacco companies are going elsewhere. Right now they're preying on the innocent and uneducated all over the world. Aggressive marketing campaigns in poor, underdeveloped countries are pulling millions of new victims into addiction.

In countries where poverty and illiteracy are the worst, more and more people are becoming smokers every year, damaging their health and hampering social and economic development. Tobacco's annual death toll is at least one million worldwide, the World Health Organization reports, adding that this number represents a minimum estimate.

Cigarette companies get substantial support from the U.S. government, which promotes cigarette exports through the federal trade representative in the Commerce Department and through export programs in the Department of Agriculture. This makes the U.S. government a global drug pusher. At the same time, the U.S. spends millions of dollars annually to fight the international traffic in illegal drugs.

The spread of smoking globally has been the work largely of American tobacco companies, says David Owen in Progressive Magazine. "Representatives from R.J. Reynolds encourage Third World farmers to switch from foodstuffs to tobacco by offering them free equipment, cheap loans, and a guaranteed market. . . . The companies have been aided substantially by the federal government, which helped export the smoking habit by including surplus tobacco in Food for Peace shipments."

Growing uneasy with such practices, some Phillip Morris stockholders, in a 1986 stockholder proposal, asked the company for a detailed report on its marketing activities in Third World countries. The proposal said: "We believe that people not sufficiently aware of health hazards related to product-use shouldn't be exploited, especially if that product-use can lead to addiction. We believe this especially applies to uninformed Third World smokers."

Naturally, the company recommended voting against this measure.

NOTHING NEW

During the eighteenth century, a cure was discovered for scurvy, a horrible disease that would sometimes ravage and kill half a ship's crew. But because the cure—fresh fruit or citrus juice—was expensive and troublesome to provide, ship owners found it convenient to remain skeptical.

After a law was passed in 1807 insuring a daily ration of lime juice for British sailors, scurvy disappeared from the navy. However, owners of merchant ships refused to accept the evidence, branding it controversial. During this "controversy," men on trade ships continued to suffer and die of this disease for another

70 years.

Though such an obstinate and callous disregard for human life may seem shocking, this is exactly what tobacco companies are exhibiting today.

THE GLOW OF SUCCESS

Meanwhile, with their profits breaking new records, the tobacco companies bask in the glow of successful free enterprise. Their glossy company reports boast of their support for arts, education, and the humanities, exuding respectability and social benevolence. The blatant hypocrisy of their posturing would be laughable, if not for the tragic consequences.

When George asks me, in class, how cigarette executives can sleep at night, I ask him, "Do you think they really care about you? You're in pain? You can't breathe? Too bad, that's your problem, George. Sure, I know you started when you were 13, but since then you've been making a free adult choice, right? Your health is not their worry. They're concerned about their private Lear jets and sending their kids to the best Ivy League schools. That's all you mean to them. Do heroin pushers worry about the misery of their customers?"

Nicotine is a drug. Most cigarette smokers are nicotine addicts. People who push cigarettes are drug pushers. This drug just happens to be legal. That's the bottom line on the multibillion dollar tobacco industry.

You may be suffering and losing your life, but don't take it personally. It's only business.

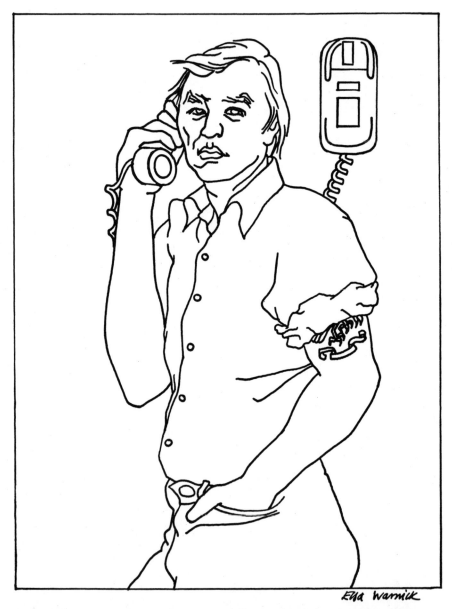

"You mean I should see a shrink?" Norman asked with contempt. "What's done is done."

CHAPTER
11

HOLDING ON
FOR DEAR LIFE

YOUR LOVE AFFAIR WITH NICOTINE:
LETTING GO

For the fortunate, life comes together when they stop smoking. For others, life seems to fall apart.

If cigarettes are so addicting, how is it that some people just put them down, walk away, and never pick them up again? Why is it others must struggle and fight, only breaking free from this love-hate relationship after repeated attempts and an enormous effort? And why do others, even when threatened with severe illness, never succeed?

While most smokers are addicted to nicotine, some are more dependent on it than others. Though the drug is the same, people are different. They have different degrees of attachment to their drug.

When I first started teaching people how to stop smoking, I assumed that those with the worst physical symptoms would have the easiest time stopping. But too often the opposite proved true.

The fact that they were still smoking, even though it had nearly destroyed their health, proved what an exceptionally powerful hold their addiction had on them.

Norman was a classic example. When he came into my seminar, he had been smoking for 26 years and had some very impressive reasons for wanting to stop. On his sign-in questionnaire he listed problems he'd been having for years: cough, phlegm, congestion, headaches, low energy, chest pains, bronchitis, shortness of breath, sinus trouble, circulation problems and emphysema.

What finally brought him in for help was having an operation in which half of his right lung had been removed. Because of this, Norman was sure he wasn't going to have any trouble staying off cigarettes, if he could just stop in the first place. The seminar helped him do that—and he thought that was all the help he needed.

He did all right for the first couple of weeks, but from then on it was touch and go. Every time I talked to him, he was on the brink of smoking. His life was in constant turmoil. He didn't get along with his boss at work. He would become enraged after a conversation with his wife, who had left him a few months earlier, taking their seven-year-old daughter with her. Norman was intensely lonely, and he was fixated on the idea that a cigarette would somehow make it all better.

Trying to help him understand his deep attachment to smoking, I asked him about his past. As a child he had lived in poverty; he suffered neglect and abuse from an alcoholic father. He was traumatized by his parents' divorce. He started smoking when he was eight.

His adult life had been just as difficult. After serving in Vietnam, he was never able to settle into a normal life. Though depressed and doubtful that his own life was worth fighting for, he wanted to stop smoking—and stay alive—for the sake of his daughter.

I told him he'd better get some counseling if he wanted to stay off smoking. "Norman, you're carrying around a lot of emotional pain from the past, and you've used cigarettes to avoid facing it since you were a child. It's time to start working through this, and you're going to need help."

"You mean I should see a shrink?"

"Maybe not a psychiatrist, but how about seeing a counselor or finding a support group so you can understand how your past has affected you? Then you can learn how to feel better about yourself in the present."

"Oh, what's done is done," he said. "It's not going to do any good to talk about it."

I kept working with him over the next several weeks. But he wouldn't consider getting extra help, even when he admitted that he was so worked up over his family problems he could hardly sleep or put in a good day's work.

It came as no surprise when he told me he'd relapsed.

"I blew it," he said. "I finally broke down and smoked."

"What happened? How did you get yourself to do that," I asked.

"Easy. My wife dropped a bomb on me the other night. She's moving out of state and taking my daughter with her. I said to myself, 'Forget it. Why bother?'"

"Okay, I understand how you were feeling. But did you try at all to answer the question you asked yourself, why you should bother? Did you try to talk back and remind yourself that you were 'bothering' because of your breathing, because smoking was killing you?"

"No, I just smoked. If my daughter's not going to be around, what difference does it make?"

Cigarettes were threatening his life, but Norman was unable to stop smoking for long. He depended on them for relief from emotional pain. Until he could overcome his low self-esteem and learn to care about himself as much as he did his daughter, he'd never be able to fight for himself. But at age 40, he just wasn't open to exploring new ways of facing his problems.

Over the years I've worked with many similar people, whose deep attachment to cigarettes is rooted in a traumatic past. Unwilling or unable to get adequate help for their problems, such people use cigarettes as a form of self-medication, as a way of numbing pain. And they usually started smoking at a very early age.

LIKE MAGIC

Of all the drugs in society, tobacco is the easiest for kids to get their hands on. Troubled youth take to nicotine like fish to water.

Smoking fills a deep psychological void, serving as a substitute for what kids really crave: love, acceptance and security. Since this drug is so addicting, kids are easily hooked. They've been primed not only by inner needs, but by the people around them.

Again and again, through the years, they've seen their parents or other adults handle stress and anxiety by lighting a cigarette. Without realizing it, they've been programmed to smoke. They've been preconditioned to use cigarettes to cope with emotional trauma.

This is exactly what happened to me. I was 17, trying my best to adjust to yet another new family. My foster parents smoked constantly. In the morning the suffocating stink would come wafting into my room and I would wake up with stuffy sinuses, struggling for breath. I hated it.

One day, faced with an impending disaster in my life, I suddenly noticed a pack of cigarettes on the table. "Maybe this will help," I thought, even though before then it had never once entered my mind to try one. I walked over, took one out and went up to my room. I sat down on my bed and lit the cigarette, inhaling that first puff as if I'd been smoking all my life.   Instantly, I felt giddy and lightheaded. I lay back on my bed, smiling. The worry dissipated like magic; I had the feeling I'd discovered something fantastic. I never experienced that euphoria again, but I was a smoker for the next 20 years.

If I'd had some other source of help, I might never have started smoking. But there was no one I felt comfortable confiding in. I had to fend for myself.

As a child, if you are neglected or abused and are unable to find solutions to pain and insecurity, you will look for a way to stop those feelings. It's normal; you're simply trying to survive.

In childhood, you have no control over what happens in the adult world around you. But once you discover nicotine, you're no longer powerless. Now you have control over the way you feel. Something nobody else was able to do for you, you can now do for yourself. You can calm yourself down, make yourself feel better when bad things are happening. You become self-sufficient. It's you and your cigarettes against the world.

PERFECT FRIENDS

As a teenager, you may feel proud and grateful to have discovered smoking. Cigarettes seem to fit the description of the perfect friend: dependable, always available when called for, a companion for every mood, occasion, or circumstance. How clever of you to have finally found this solution for yourself. Why didn't you think of this before?

While other relationships in your life have been baffling, inconsistent and frustrating, now you have a relationship you can count on. Though the people in your life have caused distress and pain, cigarettes are there to soothe and comfort you. But little do you realize you are developing a love relationship that will inevitably turn to hate.

At first, you may have very few problems with smoking. Cushioned by youthful stamina, you easily bounce back after staying out late, eating poorly or smoking too much. As long as the side effects of tobacco are temporary, you feel no need to stop using it.

But as you get older, you tend to smoke more cigarettes. At the same time, your body's ability to handle the abuse goes down. You no longer wake up fresh—cleansed and restored by a mere night's sleep.

To get going in the morning, you may need a steamy shower to clear clogged airways, aspirin to relieve a headache, two cups of coffee to get the mind working. Walking up the stairs at work, you're short of breath. In the evening, you no longer feel like going out. And on weekends, you lie around watching TV, rather than do things you used to enjoy, like swimming, playing basketball, dancing, or hiking.

Your get-up-and-go has gone up in smoke.

DENIAL

These subtle changes develop so gradually, it's easy to blame them on getting older. And if someone suggests cigarettes might be to blame, you're likely to bristle. You'll argue that you just need more vitamins or more sleep, that you're under too much stress and need a vacation, or that you have an allergy.

Why this resistance to seeing the obvious?

To admit that smoking is threatening your health and well-being undermines a deeply held belief that cigarettes are a helpmate, friend and companion, that smoking enhances the quality of your life.

Facing the truth is to be caught on the horns of a dilemma: the relationship you depend on to get through life is taking your life away.

How can you admit that cigarettes are a problem? They're your solution.

FLASH OF INSIGHT

Then something happens to break the denial.

For Vera it was a heart attack. For George, it was emphysema.

"I was a heavy smoker for years and never had any problem with it," George told me. "One winter I got a cold that turned into pneumonia, and I went into the hospital. After I got home I expected to return to my normal life, but I couldn't. When I went back to the doctor, he told me I had the beginnings of emphysema. I've never been quite the same since then, but I sure didn't want to get any worse."

For others, it was a moment of revelation. "I was always a good swimmer," Bill said. "When I visited my home town one summer and went swimming with some of my old buddies, I discovered I couldn't make it across the lake anymore. My friends still could."

Stephanie told me: "I thought my fatigue and loss of energy was just part of reaching 30, until I read an article about smoking that described all the problems I'd been having. That's when I first decided to quit."

Once the denial is broken, the logical conclusion is to stop smoking. Often that's exactly what you try to do: just put them down and walk away. After all, haven't you always said that the day smoking became a problem, that was the day you'd quit?

Unfortunately, becoming aware of the problem is not always enough to solve it. Breaking through denial—admitting that your 'friend' has become your enemy—is only the first step. To get free, you have to break your connection to the drug.

STUNTING GROWTH

Drugs numb emotions, and smoking really does "stunt growth," our emotional growth. Shooting up drugs all day neutralizes pain and discomfort, feelings that might otherwise prod us to change and to find solutions to our problems. Instead, we use cigarettes to mask the unhappiness and discomfort in our lives.

If we don't demand from life what we really want, it's certain we won't get it. And while drugs help us tolerate disappointment, they can't make us happy. Through the years we become more and more dependent on cigarettes, unaware of the role they're playing in our emotional lives. They're a consolation for what we don't have and a substitution for what we really want: respect and appreciation; love and affection; excitement and fun; rewarding work; peace of mind; a sense of meaning and purpose in life.

Having a relationship with a drug—especially an obsessive love-hate relationship—can take the place of *real* relationships: with our deepest selves, with other people and with a source of spiritual power.

In class, when Vera was talking about missing her cigarettes, I asked her what exactly she got out of smoking.

She hesitated a moment before saying, "I lose myself in a cigarette. I forget about what really bothers me."

Many people seem to put their lives on hold. They eat, drink, smoke and watch TV, stuffing and distracting themselves. At 45, overweight, out of shape, and losing their health, they know they need to change. They decide to stop smoking, expecting to feel good because they're doing the right thing for their bodies. They're disappointed and even shocked to discover how miserable they feel emotionally.

Change requires effort and risk. This is exactly what these smokers have been avoiding for years. Instead of working through problems, they've been repressing their feelings of rage, despair, resentment or depression. When they attempt to stop smoking, their emotions rush to the surface. They may feel like they're going crazy, that their whole lives are falling apart. Since the emotional pain is even greater than the physical pain smoking has caused, they're quickly overwhelmed. Regardless of how much they need to stop, smoking seems like the only way to regain control and bring things back to normal.

To keep their feelings at bay, they hang on to their addiction for dear life. And they die smoking, because they refuse to confront emotional issues or the trauma of their past.

But they *can* let go of their cigarettes if they are willing to seek help and work through their problems, maturing in the process.

While Norman wasn't willing to follow my advice, Jackie was eager to get extra help. During a follow-up call, in her first few days of withdrawal, she talked about how she felt things were going wrong in her life.

"Everything feels out of control," she said. "I've been wanting to smoke and I feel like crying all the time. I'm afraid I'm just not going to make it."

"Do you want to go back to smoking 30 cigarettes a day?"

"No, I don't. I already feel better physically."

"Then what do you want exactly?" I asked.

"I guess I just want relief from these feelings."

"Okay, you want to drug yourself because you feel bad. But do you feel so bad that you think you absolutely have to smoke? That you can't stand it?"

"Well, I guess it's not that bad."

"All right. Tell me more about these feelings."

"Well, I've been thinking about my husband, about Allen and me. I'm angry at him. I think we haven't faced a lot of things in our relationship. I'm also mad at him about me not smoking."

"What do you mean by that?"

"Remember I told you that Allen stopped smoking on his own after I took your program? I'm afraid that if I go back to smoking, he will, too. And then he'll blame me, like he did once before."

"Oh, I see. You can't smoke because Allen is depending on you to stay off so that *he* can stay off."

"I guess so."

"The first thing you need to do is to get out of deprivation. Write this down: 'I am an adult. I have choices to make. I am responsible for the consequences of my actions.' Then write the same thing for Allen."

"Don't we teach children that no matter what anybody else does, they are responsible for their own behavior?" I asked. "Aren't you two acting like little children? 'Mommy, Jackie made me do it!'"

"You're right," she said, laughing.

"If you try to lead your life on the basis of what's good for Allen, you won't live your life honestly. And the inevitable result here will be that you'll start sneaking cigarettes. Remind yourself, 'I can smoke. I don't have to stay off so Allen will'."

"Okay. I guess I have been feeling deprived. I'll watch out for that."

"Good. Now, what do you think the crying is really about?"

"Well, I've been with Allen since I was 18, and sometimes I have to face the fact that I might not want to be in this relationship anymore. And I don't know how to talk to him about it."

"Okay, this is obviously something that's going to be very difficult for you to deal with. But you don't want to make a rash decision about it, do you? What you need to do is focus on what you're doing for yourself right now, freeing yourself from a life-threatening drug addiction. That's more than enough to concentrate on for now. You'll be feeling a lot calmer soon, and then you can start working on your relationship. If you think you need it, you should get some counseling."

"All right," Jackie says. "I'll make an appointment."

"Good. And don't forget, going back to smoking won't solve problems in your relationship. You have those whether you're smoking or not."

ADDITIONAL HELP

Not everyone who stops smoking needs counseling.

But for people who have been abused and hurt in childhood or adolescence, coming back to life emotionally is painful and scary.

How am I going to handle my anger and quiet myself down? How am I going to stop those tears?

These people need a safe place where they can allow their feelings to come to the surface, where they can let themselves cry and rage and wish it had all been different—then let go of this wish and get on with their lives.

Sixty percent of the people who come through my program had traumatic experiences as children or were raised in alcoholic families. These people need counseling, self-help classes and support groups that help them accept and work through the "accidents" of childhood. They may need help from the program known as Adult Children of Alcoholics or from counselors trained in this field.

If they are coping with other addictions, they may need to take advantage of programs like Alcoholics Anonymous and Overeaters Anonymous.

DEPRESSION

Some people, despite their determination and intense desire to stop smoking, tend to relapse over and over again, not understanding why. Without realizing it, they may be suffering from depression, using nicotine as a mild euphoric to treat this condition.

Clinical or major depression is characterized by such symptoms as dramatic changes in mood, loss of interest in usual activities, insomnia, and lack of energy and ability to concentrate. It can last anywhere from two weeks to more than six months or a year.

Such depression can be an insurmountable obstacle to giving up cigarettes, and smokers who suspect they may suffer from this condition should seek medical advice. Depression can be successfully treated by drugs or therapy more than 80 percent of the time, many experts believe.

Their emotional balance restored, even the most dependent smokers will have a fighting chance to liberate themselves from their slavery to nicotine.

LETTING GO

As smokers, no matter what our differences in background, we've been using drugs to control our feelings. And recovery from addiction means temporarily letting go of that control, being willing to feel sad or angry for a while.

Stopping smoking does involve loss. But we haven't lost our cigarettes or our right to smoke them. What we have lost is the ability to smoke comfortably, without fear or pain, and we will never have that ability again.

We can grieve over that old relationship, the "perfect friend" that did only good things for us, but it's gone forever. We're no longer kids. Our bodies refuse to process, quietly and submissively, the hundreds of known toxins and carcinogens in cigarette smoke. The party's over.

For children is distress, it's normal to reach for the first comfort at hand. But as adults, if we insist on clinging to our childish solutions, the whole person suffers. We don't ever grow up. We don't get to know ourselves and learn to resolve the feelings that are causing us such pain to begin with.

As we mature, we have the ability to choose other options besides using food or drugs to numb and comfort ourselves. We don't have to go on sucking our thumbs. We can choose not to respond mindlessly and helplessly to our old impulses, as they destroy our health and our lives. Breaking our dependency on nicotine will not only improve our health, it will also teach us skills that make us more competent and independent in directing our own lives.

What makes us human is the ability to foresee the consequences of our actions and the freedom to make responsible choices.

We can say no to nicotine. But it isn't enough to "just say no." We have to be willing to say yes to change.

The best reward of teaching people how to stop smoking is sharing the joy of their success. Eight of the clients portrayed in this book were comfortable ex-smokers one year after taking the seminar.

CHAPTER
12

WHAT WINNING MEANS

ALL YOU EXPECTED—
AND MUCH MORE

Although it's been over ten years since I stopped smoking, I can still recall the distinct pleasure I felt on my first anniversary as an ex-smoker. I was walking down the street, thinking that this time it wasn't three hours or three days since my last cigarette— it was a whole year.

"You made it. You found the way out!" I told myself, feeling a surge of power and joy that seemed to lift me right off the ground.

Unlike the momentary boost of the drug, which eventually left me feeling empty and anxious, fearful for my health and my very life, the success of being free of nicotine has stayed with me. The feeling of pride and self-esteem is constant. And every time I choose to respect my life rather than give in to the addiction, I get to succeed again.

A CHANCE TO GROW

When we learn that our love for nicotine is really drug dependency, we tend to react in purely negative terms, as if it's something to be ashamed of and stamped out of our lives.

"It makes me so mad," Jackie complains to me in class. "I'd just like to be someone who never smoked, so I wouldn't have to put up with this problem."

We can either whine about the problem and end up making it that much harder to change, or we can use this opportunity to learn more about ourselves. Breaking free from an addiction is prime time for emotional growth and self-knowledge. It's a chance to become happier, more successful and confident.

Stopping smoking marks a real turning point in people's lives. No longer helpless victims of our addiction, we learn that we need not be victimized in other ways: by our jobs, our relationships, our fears, and our past. Better health also boosts our attitude and energy. We're able to use our new-found vitality and alertness to live more creatively.

Many ex-smokers have told me, "Now that I've solved this problem, which seemed so impossible, I know that if I put my mind to it, I can do just about anything."

LEARNING FROM SUCCESS

The best reward of teaching people how to stop smoking is sharing the joy of their success. All of the clients portrayed in this book, with the exception of Norman, were leading lives as comfortable ex-smokers one year after taking the seminar. Since they'd worked their way though a number of crises without smoking, none of them were worried anymore about being caught off guard and having a sudden relapse.

While they still had occasional desires to smoke, no one expressed the slightest desire to go back to smoking. They were getting too many benefits from being free of their addiction. All of them felt grateful, and at different times during their first year, they talked about the changes being an ex-smoker had made in their lives.

Lisa called me after getting her annual check up. "My doctor told me my breathing was almost back to normal, and that I would have a lot less trouble with asthma if I didn't start smoking again."

"That's great. Any other problems? Still feeling a little envious of your friends who still smoke?"

"Oh, no," she said. "In fact, I get the impression they're a little jealous of me. They don't admit it, but it's the way they ask, 'How did you manage to quit, anyway?'"

A couple of months after my last conversation with Jackie, she ran up to me in a supermarket and gave me an unexpected hug.

"I'm so glad I took your advice and went back to counseling," she said. "I never realized I had so much anger in me until I stopped smoking. It's amazing, but I hardly think of smoking anymore now that I'm feeling better about my relationship and can see some hope."

"Well, you sure seem a lot happier."

"Oh, I have so much more energy, I can't believe I used to sit around all day eating junk food and smoking cigarettes."

When people stop smoking successfully, they feel an enormous sense of relief and gratitude. And like many of my other clients, both Mike and Stephanie wrote to express their thanks.

"I've got a lot more energy," Mike said in his letter. "I've been out there throwing a football with my son. And when I talk to him about smoking dope, I don't feel like a hypocrite anymore."

Stephanie was enjoying her new job as a television reporter. "But I can just imagine what it'd be like if I were still smoking. I'd be stressed out and staying up half the night smoking and worrying about my next deadline."

For Vera, stopping smoking meant not only better health and a different lifestyle, but a change in her whole attitude.

I talked to her after her vacation to Reno. She had made it back without smoking, but she sounded very unhappy.

"What's the matter, Vera? Didn't you have a good trip?"

"Oh, it was a great trip. Everything you told me to do worked beautifully. But since I've been back, the desire to smoke has been stronger than ever."

"What's going on?"

"Well, while I was gone, my ex-husband's wife died, and I just found out my kids are going to stay with him for a week."

"Why should that bother you, Vera?"

"Listen. I worked for years and helped put my husband through business school and then after 20 years he just took off. I finished raising the kids, and I'm the one who was always there for them. Now when he says he needs them, they come running."

"And that's making you want to smoke?"

"Oh, yes. I got in an argument with my daughter this morning, and I was just so close to smoking."

"You sound pretty angry. Is that what's behind your desire to smoke?"

"Yes, I'm furious. How can they forget what he did to me?"

"Maybe you have unreal expectations, Vera. Do you think because your kids love you, they should automatically hate their father."

"Sure, after what he did to me."

"Well, I'm sure they don't like the way he treated you years ago, but why should that destroy their love for him? They have their own relationship to their father which has nothing to do with the way they feel about you. They have a natural bond with him that you can't change. Do you see what I'm saying?"

"I've never really thought about it that way."

The next time we talked, Vera told me, "I've been thinking about what you said. And you know what I realized? I have been an angry, bitter, resentful woman for 15 years. I've been using tranquilizers, sleeping pills, cigarettes, and all I've been doing is holding the feelings in. Maybe I need to work on letting go of this stuff from the past—it isn't helping anyone."

Once Vera began taking responsibility for her own feelings, she was less frustrated and angry, and her desires to smoke diminished. She saw that although she couldn't control what her children and ex-husband did, she could do something about her own feelings.

Stopping smoking for Vera served as the catalyst for a profound insight into the way she felt about herself and other people.

Bill had a similar experience. Not smoking had been relatively easy for him for the first few months, and he always seemed the picture of a confident young attorney who knew exactly where he was going. But then he revealed a much different side of himself.

It was late in the evening when he called me from his office, where he'd been working on one of the toughest cases of his career.

"The whole firm is watching to see how I handle this thing," he said. "I feel like it could literally make or break me, and I'm beginning to get a little panicky. I'm really worried that I'm not going to win—and I sure want to smoke."

"How have you been responding to these desires, Bill? Are you working through them?"

"Yes, I am, but it doesn't seem to be doing any good."

"You mean you're smoking?"

"Of course not."

"Then it *is* doing some good. So keep working on it. Now tell me, how is your mind trying to get you to go ahead and give in?"

"Oh, I guess I get to the point where I just don't care. I just want to smoke."

"Okay, let's look at that a little more closely. It's been about four months since you stopped. How much of that time do you think that you've felt this way, that you just don't care."

Bill paused for a moment, then laughed. "Well, probably only the last three hours."

"Good. So the rest of the time, when you've been playing racquetball or working with your clients, have you been feeling good about not smoking?"

"Oh, absolutely."

"So it's only when you get into a rough spot and want to drug yourself through it, that you start telling.yourself 'I don't care.' Right, Bill?"

"Well, I sure feel like I don't care."

"Then it's critical that you make the distinction between 'feels like' and reality. You obviously do care or you wouldn't have stopped smoking, and you wouldn't be struggling with it now. Does that make sense?"

"Yeah, I really don't want to go back to smoking."

"Okay. Sometimes you may wish you didn't care, but you have ideals and standards of health that make it impossible for you to be a happy smoker. So why do you think you're trying to convince yourself that you don't care? Can you describe the feeling that's behind all this worrying about work? Is it anger? Resentment?"

"Well, no. It seems more like a heavy feeling, a sadness. It's the way I used to feel around my Dad when I was a kid, like he never really approved of me or had much faith in me. I still get the impression he's just waiting for me to screw up."

"Do you believe that?"

"No, there's no real reason to think that. I know he really cares about me; he just doesn't know how to show it. Maybe that's why I've always had feelings of not being quite good enough at anything I did."

"And maybe that's why you're having these strong desires to smoke," I said. "If you tell yourself you don't care, you can go ahead and drug yourself and suppress your feelings. Of course, if you do that, then you have to go on drugging yourself, day in and day out. What you need to do now is to separate these childhood feelings from your reality today. They're just feelings, and they're temporary, so you don't have to kill yourself over them."

Bill got through that night, and he went on to win his case. Later he told me what an impact this conversation had on his thinking.

"It made me aware of a problem I wasn't even conscious of before. It had to do with a basic feeling about myself, that somehow I was inadequate because I didn't measure up to my Dad's standards. Now I see how absurd that was, and there's been a kind of subtle shift in my whole attitude. I'm a lot more relaxed and comfortable with myself."

Like Bill, most people stop smoking to improve their health. But they often discover benefits that go much deeper than that.

Ellen stopped by my office one day to tell me how much being free of nicotine had affected her work.

"I always prided myself on being a first-rate counselor," she said. "But now it's clear that smoking created a kind of screen between me and my clients."

"A kind of emotional smokescreen," I said.

"Yes. Drugs numb your feelings. So you're just not as fully alive as you could be, you're not fully there."

"Cigarettes tend to cut you off from other people, don't they?"

"And even from your own self," she said, "because the relationship with your drug comes first. I would never have admitted this while I was smoking, but I was really unhappy with myself, preoccupied, knowing I was doing something terrible to my health."

Over the last ten years, I've had the pleasure of seeing many hundreds of people successfully end their love-hate relationship with nicotine. Like the clients who appear in this book, they came in to stop smoking only to discover that this was the first step in transforming their lives.

They learned that, instead of losing something, they were letting something go. And rather than thinking of stopping smoking as an end, they realized that it was really a beginning.

In Brief

Key Points for Success

❖ SET A DEFINITE TIME TO STOP.

What doesn't stop goes on. While there's never a perfect time to give up an addiction, choose a time to stop when you can make this effort your top priority.

❖ DON'T TRY TO GET RID OF DESIRES TO SMOKE.

The discomfort of wanting to smoke is temporary and will get rid of itself. When you have a desire to smoke, work through the five steps from Chapter 5.

❖ DON'T SUBSTITUTE FOOD.

If you "smoke" food whenever you have a desire for a cigarette, you will not only gain weight, you will never break your addiction. Desires to smoke will continue to nag you until you finally break down and give in.

❖ STAY CLEAR OF FEELINGS OF DEPRIVATION.

Telling yourself you can't smoke is a lie and will make you feel so miserable you will run back to smoking. Remember that you *can* smoke. You just can't do it the way you'd like to: now and then or without damaging your health.

❖ GIVE UP THE ILLUSION OF HAVING "JUST ONE."

One puff or one cigarette has never been enough for you, and it never will be enough. It will inevitably take you back to smoking your normal amount. The crux of beating a drug addiction is knowing this: it's the first one that does you in.

❖ CHOOSE BETWEEN *REAL* OPTIONS.

The only real options you have are these: going back to smoking with all the terrible consequences or staying off smoking with the many benefits. You don't have to *like* this reality, but you'd better accept it.

❖ FOCUS ON BENEFITS—CONTINUALLY.

Keep in mind the specific benefits you are gaining from being free from your addiction. Counter your compulsion to smoke by remembering what you want more: your breathing, your freedom, and your peace of mind.

❖ GET SMART ABOUT JUNKIE THINKING.

Every time you have a junkie thought, identify it and talk back to it. If you do, these irrational thoughts and plans will eventually lose their power over you.

❖ TAKE TIME OUT WHEN YOU HAVE A DESIRE TO SMOKE.

During withdrawal or in any high risk situation, get away by yourself for a few minutes to review the five steps and get your thinking back on track.

❖ BE UNCOMFORTABLE—GRACIOUSLY, AND ON PURPOSE.

The discomfort caused by wanting to smoke is temporary and harmless, and it's your means to escape from slavery. The desire to smoke will gradually become less intense and less frequent until most of the time you will feel like a non-smoker.

❖ PREPARE FOR HIGH-RISK SITUATIONS.

Most people who relapse do so within the first three months, because they are not prepared for such things as traveling or emotional upsets. Stay alert and beware of overconfidence.

❖ YOU DON'T HAVE TO CHANGE YOUR LIFE.

Drinking coffee, having a glass of wine or eating spicey foods will not make you smoke. They can make you *want* to smoke. So your job is to *treat* the desires to smoke rather than avoid them. Change your thinking, not your daily activities.

❖ USE DREAMS CONSTRUCTIVELY.

Dreams about smoking are very common and do not mean you are doomed to relapse. The anxiety you feel in a dream when you realize you're smoking will teach you that, although you can smoke, you will never be happy with it.

❖ REMEMBER, THERE'S NO CURE FOR ADDICTION.

You will never be a *non-smoker*. A non-smoker is someone who has never had a problem with smoking, has never struggled to take control of that problem, and never has to worry about losing control. You're an ex-smoker, and although you can be a confident and relaxed ex-smoker, you are always susceptible to relapse.

❖ EXPECT TO HAVE A THREE-MONTH FLARE-UP.

Many ex-smokers relapse toward the end of the third month, because their health has improved and the side-effects of smoking have disappeared. Read your 'remember' lettter and don't imagine that time has cured your addiction. One puff and you will soon be back to smoking compulsively.

❖ **DON'T NAG OR PREACH AT OTHER SMOKERS.**

You're only one puff away from a pack a day yourself. Take care of your own recovery and watch out for the influence of other smokers around you. Don't glamorize smoking; remember what it was really like to have to smoke all day every day.

❖ **GET EXTRA HELP IF YOU NEED IT.**

Attend a self-help group or see a counselor to work through feelings you've been drugging away all these years. Letting go of these feelings from the past and learning new ways to cope with the present will help you become a happier, more comfortable ex-smoker.

AFTERWORD

Adult Choice
or Child Seduction?

It's so commonplace we hardly take notice: a group of teen-agers, often as young as twelve or thirteen, hanging out in the park or on the street corner, chatting happily and puffing away on their cigarettes.

It's easy to wink at this behavior, as most people do, including the media. "They're just kids experimenting," they say. "It's harmless rebellion." Recently I heard a caller on a talk show complain, "Why is the government messing around with a flimsy issue like kids smoking, when there are so many real problems out there?" The talk show host vigorously agreed.

People are missing the connection. This issue is not just about kids smoking. It's about a deliberate campaign by the tobacco companies to get young people hooked on a dangerous drug—one known to cause disability and early death.

The billion dollar advertising campaigns, using camels and cowboys to appeal to youth, definitely work. Three thousand kids become new addicts—customers—every day, and one third of them will suffer and eventually die of this addiction.

That adds up to more than 400,000 victims a year in the U.S.

Is this a flimsy issue?

Youthful rebellion is not the point. A kid with a cigarette will eventually be an adult on a respirator, an adult with a heart attack or lung cancer—that's the point.

Defending their "rights," the tobacco industry claims that smoking is an adult choice, that we don't need government bureaucrats telling us what to do. This is a ridiculous argument. Ninety per cent of smokers start between the ages of 11 and 18, surveys reveal. Teenagers who start smoking today will suffer many years of anguish and pain as adults before they stop—if they're lucky enough to stop at all. By that time, they will have spent some $20,000 to $30,000 on tobacco.

Although the cigarette companies deny that they market tobacco products to youth, their own records tell a different story. One R.J. Reynolds memo, according to the Washington post, suggested that a new cigarette brand could be marketed as a way to cope with the pressure of being a teenager.

In any case, if the makers of cigarettes actually refrained from targeting kids with their ads, they would eventually put themselves out of business. Without fresh new recruits, who would replace the smokers who die off?

"The R.J. Reynolds Tobacco Company has the gall to blame the problem of youth smoking on everyone but itself," said one industry watchdog, the Coalition on Smoking or Health.

The American Heart Association said the company has been marketing to teenagers for 30 years. "They've been robbing the cradle of America's kids for too long," said an association spokesman. "How can anyone believe they don't market to kids when their own documents prove it?"

Even the American Medical Association, which failed to speak out forcefully on the subject for so many years, now calls smoking a pediatric disease, accusing the tobacco companies of preying on children.

They're right. The compulsion to smoke is like a contagious disease we 'catch' as children. The last thing we imagine is that this condition will be part of our lives for as long as we live. In a real sense, this disease is incurable. Even if we finally manage to stop giving in to the craving to smoke, we're always susceptible to relapse.

By allowing cigarette companies free reign to market their products, isn't society giving this industry permission to infect as many young people as it can with a kind of deadly virus?

What could be wrong with curtailing this kind of predatory marketing?

Even with strict new controls to keep cigarettes out of the hands of youth, education is still needed to help prevent kids from smoking. In the past, such education has usually emphasized the dangers of smoking to one's health. For most young people, what will happen to their health in 20 or 30 years is far too remote from their present lives to serve as an effective deterrent.

Almost universally, kids believe that if they do start smoking, they'll be able to stop before their health is affected. What they need to learn is that smoking—this seemingly harmless rite of passage into adulthood—is almost certain to subject them to decades of slavery and regret. To their dismay, most of them won't be able to stop just because they want to, just because smoking has now become inconvenient and worrysome. They'll have to endure years of pain and fear before they can finally face up to drug withdrawal.

Preventive education needs to dramatize and explain the addictive power of nicotine and the insidious grip it has over people. This is no easy task, and far more work needs to be done to develop effective methods for teaching young people what it really means to be an addict. An imaginative and sophisticated approach will open young people's eyes, so that they can avoid wandering blindly into the quagmire of an unnatural and degrading dependency.

THE AUTHORS

Patricia Allison is a leading authority on nicotine addiction. For over 20 years she has taught thousands of people to stop smoking. For many years, her Stop Smoking Program was offered in 16 medical centers in Oregon and Washington. An expert at helping the chronic smoker succeed through individual step-by-step coaching and training, she writes from many years of personal and professional experience.

A dynamic and forceful speaker, Allison brings fresh insights to a problem too often treated as a naughty bad habit. With her passionate concern for individuals trapped without hope in a deadly addiction and for children seduced into smoking by crafty marketing, Allison is as inspiring as she is informative.

To book Allison as a speaker, contact BridgeCity Books at 503-220-4181. For information about Allison's new national program, see the web page: www.stopsmoking.com

Jack Yost, a writer, publisher and teacher in the field of peace education, is the author of *Planet Champions, Adventures in Saving the World.* He has worked at the United Nations as the director of the international office of the World Federalist Movement in New York and currently serves as director of the Oregon Peace Studies Association.

Books & Resources

The following selection of books has been very helpful to me, my staff and other recovering ex-smokers, and I highly recommend them.

Bailey, Covert. **Fit or Fat.** Boston: Houghton-Mifflin Company: 1977

Bailey, Covert. **The Fit or Fat Woman.** Boston: Houghton-Mifflin Company: 1989

Beattie, Melody. **Codependent No More.** New York: Harper-Hazelden, 1987

Bradshaw, John. **Healing The Shame That Binds You.** Deerfield Beach: Health Communications, 1988

Branden, Nathaniel. **The Psychology of Self-Esteem.** Toronto: Bantam, 1978

Brister, David & Phyllis. **The Vicious Circle Phenomenon.** Birmingham: Diadem Publishing, 1987

Burns, David D. **Feeling Good.** New York: Wm. Morrow & Co., Inc., 1980

Ellis, Albert. **A New Guide to Rational Living.** North Hollywood: Wilshire Books, 1975

Emery, Stewart. **Actualizations.** Garden City: Dolphin Books, 1978

Goldhor Lerner, Harriet. **The Dance of Anger.** New York: Harper & Row, 1985

Halpern, Howard. **How To Break Your Addiction to a Person.**
New York: McGraw-Hill, 1985

Halpern, Howard. **Cutting Loose.** Toronto: Bantam/Simon & Schuster,
1976

Peele, Stanton. **Love and Addiction** New York: Taplinger, 1975

Norwood, Robin. **Women Who Love Too Much.** New York: St.
Martin's Press, 1985

Smith, Manuel. **When I Say No, I Feel Guilty.** New York: Bantam
Books.

Roth, Geneen. **Breaking Free From Compulsive Eating.** New
York: Bobbs-Merrill Co., 1984

Washton, Arnold & Boundy, Donna. **Willpower's Not Enough.** New
York: Harper & Row, 1989

Whitfield, Charles. **Healing The Child Within.** Deerfield Beach:
Health Communications, 1989

Wilson Schaef, Anne. **When Society Becomes An Addict.** New
York: Harper & Row, 1987

Woititz, Janet. **Adult Children of Alcoholics.** Deerfield Beach:
Health Communications

As an ex-smoker, you may discover that there are emotional issues or conditions in your life that you need to address. It is quite likely that there is a self-help group or a community-sponsored service that will benefit you.

A first step may be to call your local library and ask if there is a publication available that lists community services. This publication may also include a list of counseling services. Hospitals often have a variety of workshops at reasonable fees and available during evening hours.

The organization, **Alcoholics Anonymous,** listed in the phone book in almost every city , can direct you to groups such as:

Al-Anon — A group for families and friends of alcoholics and drug abusers.

OA — (OverEaters Anonymous) A group for compulsive overeaters.

Women/Men Who Love Too Much — These meetings help those with dysfunctional and co-dependent relationships.

ACAP & ACOA (Adult Children of Alcoholic Parents)—This group is excellent for adults who were raised in alcoholic or otherwise neglectful or abusive families and have difficulty with low self-esteem and interpersonal relationships.

There are many 'anonymous' groups, and in order to discover which ones meet your needs, attend several and ask for schedules and information regarding other groups. One major benefit will be to discover *that you are not alone.*

Index

INDEX

INDEX

WORKSHEET

Write your Remember Letter here.

WORKSHEET

Write the five steps here.

172

WORKSHEET

Write the high risk situations and emotions that
you need to be most concerned about.

WORKSHEET

Write your list of benefits here.

WORKSHEET

Write your Remember Letter here.

ENDORSEMENTS FOR PATRICIA ALLISON & HER BOOK

"Hooked—But Not Helpless is a well-written, exceptionally clear book. A hard-headed, truly no-nonsense approach."
— Albert Ellis, Ph. D.
Author, *A New Guide to Rational Living*

"In these pages, Patricia Allison aims a blast of clarity, reality and hope at the deadly business of tobacco addiction. A brightly positive attitude pervades her presentation of a strategy that has proven successful in helping thousands to stop smoking."
— John Nance
Author, *The Gentle Tasaday*

"This book is extremely interesting, practical and useful even to the non-smoker for its insights into behavior.
— Covert Bailey
Author, *Fit or Fat*

"Thank you for your wonderful appearance on our program recently. You were great—and our audience loved you. Know, too, we'll be in touch with you again in a few months to ask you for a repeat performance."
— Duane Patterson, Producer,
Warren Duffy's "Alive Across America"

"Jerry, I'm so glad you have Patricia Allison on the air today, because this is the first person I've heard talk about smoking that makes any sense!"
— Caller
The Jerry Williams Show, Boston